Introduction to Homotopy Theory

Introduction to Homotopy Theory

Editor

Aneta Hajek

Introduction to Homotopy Theory

Edited by **Aneta Hajek**

Printed in 2017

ISBN: 978-1-68117-185-2

Library of Congress Control Number: 2015949130

Notice

Preface

Homotopy theory, which is the main part of algebraic topology, studies topological objects up to homotopy equivalence. Homotopy equivalence is weaker relations than topological equivalence, i.e., homotopy classes of spaces are larger than homeomorphism classes. Even though the ultimate goal of topology is to classify various classes of topological spaces up to a homeomorphism, in algebraic topology, homotopy equivalence plays a more important role than homeomorphism, essentially because the basic tools of algebraic topology (homology and homotopy groups) are invariant with respect to homotopy equivalence, and do not distinguish topologically nonequivalent, but homotopic objects.

The idea of homotopy can be turned into a formal category of category theory. The homotopy category is the category whose objects are topological spaces, and whose morphisms are homotopy equivalence classes of continuous maps. Two topological spaces X and Y are isomorphic in this category if and only if they are homotopy-equivalent. Then a functor on the category of topological spaces is homotopy invariant if it can be expressed as a functor on the homotopy category.

Based on the concept of the homotopy, computation methods for algebraic and differential equations have been developed. The methods for algebraic equations include the homotopy continuation method and the continuation method. The methods for differential equations include the homotopy analysis method.

In practice, there are technical difficulties in using homotopies with certain spaces. Algebraic topologists work with compactly generated spaces, CW complexes, or spectra. This book deals with homotopy theory, one of the main branches of algebraic topology.

Contents

A Discrete Model of O(2)-homotopy Theory

Jan Spaliński
Faculty of Mathematics and Information Science,
Warsaw University of Technology,
Pl. Politechniki 1, 00-661 Warsaw, Poland

ABSTRACT

A. Blumberg has shown that the category of triples consisting of a simplicial set, a cyclic set and appropriate compatibility data serves as a discrete model for the homotopy theory of spaces with an S^1-action. We show that the category of triples consisting of a $\Delta(\mathbb{Z}/2)$-set, dihedral set and appropriate compatibility data serves as a model for the homotopy theory of O(2)-spaces.

INTRODUCTION

In the recent article [1], A. Blumberg succeeds in constructing a combinatorial model which completely describes the homotopy theory of S^1-spaces. This is the final step in an area of research which started a quarter of century ago with A. Connes' paper [2]. The aim of that paper is to give a derived functor approach to cyclic cohomology. In order to achieve this, Connes introduces the cyclic category Λ, which is a modification of the simplicial category Δ. It is soon realized that the category Λ can be used to provide a discrete model for S^1-spaces, see W. Dwyer, M. Hopkins, and D. Kan [3]. More specifically, a map of cyclic sets (i.e. functors $\Lambda^{op} \rightarrow Sets$) is called a weak equivalence if it is a weak equivalence of the underlying simplicial sets. The homotopy category of cyclic sets with this structure is equivalent to the homotopy category of topological spaces with the action of S^1, where a map is a weak equivalence if it is a weak homotopy equivalence of the underlying spaces. There is a more rigid structure on the category of spaces with an action of a compact Lie group G obtained by taking as weak equivalences the maps which induce weak homotopy equivalences on the fixed point sets of all closed subgroups of G. M. Bökstedt, W.C. Hsiang and I. Madsen [4] show that cyclic sets do in fact capture the fixed point data of finite subgroups of S^1. These ideas are given a precise homotopy-theoretic interpretation in [5]. As the fixed point set of S^1 acting on the realization of a cyclic set is always discrete, it is not possible to take into account the fixed point data for all of S^1. Blumberg overcomes this deficiency of cyclic sets by introducing a category of triples: a simplicial set, a cyclic set, and appropriate compatibility data. He gives this category a model structure whose homotopy category is equivalent to the homotopy category of S^1-spaces where a map is a weak equivalence if it induces weak equivalences of spaces on the fixed point sets of all closed subgroups of S^1 (i.e. the finite subgroups and S^1 itself).

In [6], J.-L. Loday introduces the category S^d of dihedral sets (these are simplicial sets with the actions of the dihedral group of order $2(n+1)$ in dimension n). Z. Fiedorowicz and Loday show in [7] that these objects can serve as models for O(2)-spaces, the detection of fixed point sets is not addressed. As is shown in [8], the fixed points of finite subgroups of O(2) are in fact detected by dihedral sets.

In this paper we give, using Blumberg's paper as a guide, a combinatorial model for the homotopy theory of O(2)-spaces taking into account the fixed point sets of all closed subgroups.

The situation is a bit more complicated than the case of S^1-spaces, because O(2) has two infinite closed subgroups: S^1 and O(2) itself. We show that an O(2)-space can be represented by a triple consisting of a $\Delta(Z/2)$-set (defined in [7]), a dihedral set, and an appropriate compatibility map.

OUTLINE OF THE PAPER

In Section 3 we establish a combinatorial model for the equivariant homotopy theory of $\mathbb{Z}/2$-spaces. This is the category of $\Delta(\mathbb{Z}/2)$-sets, which we denote by $S^{\mathbb{Z}/2}$. Essentially, a $\Delta(\mathbb{Z}/2)$-set is a simplicial set with a $\mathbb{Z}/2$-action in each dimension, whose realization is a $\mathbb{Z}/2$-space. Some care is needed here, as things need to be arranged in such a way that there is a natural forgetful functor from the category of dihedral sets to the category of $\Delta(\mathbb{Z}/2)^{-\text{sets}}$.

In Section 4 we construct a functor $\nabla : S^{\mathbb{Z}/2} \to S^d$, whose role is to "glue" the fixed point data for the two infinite closed subgroups of O(2) onto the dihedral set. We also introduce the coupled model category $S^{\mathbb{Z}/2}{}_\nabla S^d$. This is a category consisting of triples $A=(A_{\mathbb{Z}/2}, A_d, \xi : \nabla A_{\mathbb{Z}/2} \to A_d)$. For this construction to work, we need to check that ∇ is Reedy admissible (Definition 1.6 in [1]).

In Section 5 we recall the relevant model category structure on the category of O(2)-spaces and make the final step by giving a pair of adjoint functors relating this category and the category $S^{\mathbb{Z}/2}{}_\nabla S^d$ and showing that they satisfy the assumptions of D. Quillen's equivalence theorem ([9], Chapter I, Theorem 3). We apply a slightly simplified version of Quillen's theorem which appears as Theorem 9.7 in [10].

EQUIVARIANT HOMOTOPY THEORY OF Z/2-SPACES

The first ingredient of our model of O(2)-spaces is the model for the fixed point sets of the two infinite closed subgroups of O(2), i.e. O(2) and S^1. If X is an arbitrary O(2)-space, then the fixed point set of S^1 has a natural $\mathbb{Z}/2$-structure, since $O(2)/S^1 \simeq \mathbb{Z}/2$.

The next result follows from Theorem 2.2 in Dwyer and Kan [11].

Proposition 3.1

The category $\text{Top}^{Z/2}$ of topological spaces with a Z/2-action has a model structure in which a map $f:X \to Y$ in $\text{Top}^{Z/2}$ is a weak equivalence (fibration) if both the map f and the map $f^{Z/2}:X^{Z/2} \to Y^{Z/2}$ are weak homotopy equivalences (fibrations) of topological spaces.

A discrete model for $\text{Top}^{Z/2}$ is provided by the $\Delta(Z/2)$-sets of Fiedorowicz and Loday [7], Example 2, p. 59. The category $(\Delta(Z/2))^{\text{op}}$ is the subcategory of $\left(\wedge^d\right)^{\text{op}}$ of dihedral operators (Definition 1.1 in [8]) generated by Δ^{op} and the morphisms $_{wn+1}:[n] \to [n]$, $n \geq 0$. A $\Delta(Z/2)$-set is a functor $(\Delta(Z/2))^{\text{op}} \to \text{Sets}$ and a morphism is a natural transformation of such functors. We denote this category by $S^{Z/2}$. The categories Δ^{op} and $\Delta(Z/2)^{\text{op}}$ are related by the obvious inclusion functor $j:\Delta^{\text{op}} \to \Delta(Z/2)^{\text{op}}$.

We also need the standard $\Delta(Z/2)$-sets, which are defined in the usual way: $\Delta(Z/2)\,[n] =_{\text{hom}(\Delta}(Z_{/2))\text{op}}([n],-)$. One can check that $|j^*\Delta(Z/2)[n]| = Z/2 \times \Delta^n$; we denote this Z/2-space by $\Delta^n_{Z/2}$. The collection of these forms a cosimplicial space which we denote by $\Delta^*_{Z/2}$. There is a pair of adjoint functors:

$$|?|_{Z/2} : \mathbf{S}^{Z/2} \leftrightarrow \mathbf{Top}^{Z/2} : S^*_{Z/2} \tag{3.2}$$

defined as follows:

$$|X|_{Z/2} = X \otimes_{\Delta(Z/2)^{op}} \Delta^*_{Z/2}$$

and

$$S^*_{Z/2}(X)_n = \text{Hom}_{\mathbf{Top}^{Z/2}}(\Delta^n_{Z/2}, X).$$

We need another description of the realization functor.

Lemma 3.3

There is a natural isomorphism of functors

$$|j^*(?)| \simeq F \cdot |?|_{\mathbb{Z}/2} : \mathbf{S}^{\mathbb{Z}/2} \to \mathbf{Top},$$

where $F : \mathbf{Top}^{\mathbb{Z}/2} \to \mathbf{Top}$ is the forgetful functor.

Proof
We let $\mathbb{Z}/2 = \pm 1$ with the obvious group structure. Consider the following diagram:

$$
\begin{array}{ccc}
\coprod_{n \geq 0} X_n \times \Delta^n & \xrightarrow{\ q\ } & \coprod_{n \geq 0} X_n \times \Delta^n_{\mathbb{Z}/2} \\
\downarrow{\scriptstyle p_1} & & \downarrow{\scriptstyle p_2} \\
|j^*(X)| & & |X|_{\mathbb{Z}/2}
\end{array}
$$

where $q(x,u) = (x, u \times 1)$ and the vertical maps are the canonical maps onto the quotient. It is not hard to show that if we have two elements in $\coprod_{n \geq 0} X_n \times \Delta^n$ which map to the same element in $|j^*(X)|$, then they also map to the same element in $|X|_{\mathbb{Z}/2}$. This implies that there is a well-defined map $h : |j^*(X)| \to |X|_{\mathbb{Z}/2}$ induced by q.

It is clear that every element $[x, u \times 1]$ is in the image of h. To see that elements of the form $[x, u \times -1]$ are also in the image, note that $(x, u \times -1)$ maps to the same element as $(w_{n+1} x, u \times 1)$ in $|X|_{\mathbb{Z}/2}$, where n is the dimension of x. Similar arguments lead to the observation that h is one to one: if $h[x,u] = h[y,v]$, then the pairs $(x, u \times 1)$ and $(y, v \times 1)$ are related by an element of $\varphi \in \Delta(\mathbb{Z}/2)$. However, since the last term in both pairs equals $+1$, in fact $\varphi \in \Delta$, which implies $[x,u] = [y,v]$.

Recall that the Segal–Quillen 2-fold subdivision of a simplicial set is defined by $sq(X)_n = X_{2n+1}, \overline{d}_i = d_i.d_{2n+1-i}$ and $\overline{s}_i = s_{2n-i}.s_i$. Moreover, using the ideas from Section 2 in [8], we define a functor $\Gamma: S^{\mathbb{Z}/2} \to S, X \to sq(X)^{\mathbb{Z}/2}$, where the action w of $\mathbb{Z}/2$ on $sq(X)$ is given by $wx = w_{n+2}x$ for $x \in sq(X)_n = X_{2n+1}$. It is not hard to check that $|\Gamma(X)| \simeq (|X|_{\mathbb{Z}/2})^{\mathbb{Z}/2}$. We can apply Theorem 2.2 in [5] to give the category $S^{\mathbb{Z}/2}$ the following model structure:

Proposition 3.4

The category $S^{\mathbb{Z}/2}$ has a model structure in which a map f:X→Yof Δ($\mathbb{Z}/2$)-sets is a weak equivalence (fibration) if both the map j*f and the map Γ(f) are weak equivalences (fibrations) of simplicial sets.

The pair of adjoint functors (3.2) satisfies the assumptions of Quillen's equivalence theorem. We briefly indicate why the condition in part (2) is satisfied. Suppose that the map $f: X \to s^*_{\mathbb{Z}/2}(Y)$ is a weak equivalence in $S^{\mathbb{Z}/2}$. Hence the map $j^*(f): j^*(X) \to j^* s^*_{\mathbb{Z}/2}(Y)$ is a weak equivalence of simplicial sets. However, we have $j^* s^*_{\mathbb{Z}/2} \simeq S^*$, and the usual adjunction between simplicial sets and topological spaces (May [12]) implies that the adjoint $|j^*X| \to Y$ is a weak equivalence of spaces. Next, since $\Gamma(f): \Gamma(X) \to \Gamma s^*_{\mathbb{Z}/2}(Y)$ is a weak equivalence, so is the realization $|\Gamma(f)|: |\Gamma(X)| \to |\Gamma s^*_{\mathbb{Z}/2}(Y)|$.. Since $|\Gamma(X)| \simeq (|X|_{\mathbb{Z}/2})^{\mathbb{Z}/2}$ and $|\Gamma s^*_{\mathbb{Z}/2}(Y)| \simeq (|s^*_{\mathbb{Z}/2}(Y)|_{\mathbb{Z}/2})^{\mathbb{Z}/2} \simeq Y^{\mathbb{Z}/2}$, we see that the adjoint map$(|X|_{\mathbb{Z}/2})^{\mathbb{Z}/2} \to Y^{\mathbb{Z}/2}$ is also a weak equivalence. Hence Δ($\mathbb{Z}/2$)-sets give a combinatorial model for $\mathbb{Z}/2$-equivariant homotopy theory.

THE COMPATIBILITY FUNCTOR ∇ AND THE COUPLED MODEL STRUCTURE

By Theorem 5.7 in [7], there exists a pair of adjoints

$$|?|_d : \mathbf{S}^d \leftrightarrow \mathbf{Top}^{O(2)} : S_d^*.$$

We define the cosimplicial dihedral set ∇_d^* by $\nabla_d^n = S_d^*\big(\Delta_{\mathbb{Z}/2}^n\big)$, that is ∇_d in dimension n is the dihedral singular complex of the realization of the standard $\Delta(\mathbb{Z}/2)$ n-simplex, which is viewed as an O(2)-space. More precisely, the canonical map $\varrho\colon O(2) \to O(2)/S^1 \cong \mathbb{Z}/2$ gives a functor $\mathbf{Top}^{\mathbb{Z}/2} \to \mathbf{Top}^{O(2)}$. Now we can define the main object of this section.

Definition 4.1

Let $\nabla\colon \mathbf{S}^{\mathbb{Z}/2} \to \mathbf{S}^d$ be the functor defined by

$$\nabla(X) = X \otimes_{(\Delta\mathbb{Z}/2)^{op}} \nabla_d^*.$$

The key property of ∇ is given by

Lemma 4.2

There exists a natural map $\zeta\colon |\nabla(X)|_d \to |X|_{\mathbb{Z}/2}$. This map induces weak homotopy equivalences on the fixed point sets of all finite subgroups of O(2).

Proof

We need to know that the map $\gamma_n\colon \big|S_d^*\big(\Delta_{\mathbb{Z}/2}^n\big)\big|_d \to \Delta_{\mathbb{Z}/2}^n$ is a weak equivalence upon passage to all the fixed point sets of finite subgroups of O(2). By Lemma 4.12 in [8] this is true for any O(2)-space, hence in particular for $\Delta_{\mathbb{Z}/2}^n$, viewed as an $O(2)$-space. The proof now proceeds along the lines of the proof of Lemma 4.3 in [1]: we define ζ to be the following composite:

$$|\nabla X|_d = (X \otimes_{(\Delta\mathbb{Z}/2)^{op}} \nabla_d^*) \otimes_{(\Lambda^d)^{op}} \Lambda_d^*$$
$$= X \otimes_{(\Delta\mathbb{Z}/2)^{op}} (\nabla_d^* \otimes_{(\Lambda^d)^{op}} \Lambda_d^*)$$
$$= X \otimes_{(\Delta\mathbb{Z}/2)^{op}} |\nabla_d^*|_d \stackrel{\gamma_*}{\mapsto} X \otimes_{(\Delta\mathbb{Z}/2)^{op}} \Delta_{\mathbb{Z}/2}^* = |X|_{\mathbb{Z}/2}.$$

Note that in the second line of the calculation above, in the second tensor product, it is the dihedral rather than the cosimplicial structure of ∇_d^* that is used.

Next, we recall how to merge two model categories related by a functor into one.

Definition 4.3 Blumberg [1]

Let C and D be categories, and $F: C \to D$ a functor. The objects of $C_F D$ are triples $(A, B, FA \to B)$, where A is an object of C, B is an object of D and morphisms are pairs of maps $\alpha: A \to A'$ and $\beta: B \to B'$ such that the two possible maps $FA \to B'$ are equal.

Theorem 4.4 Blumberg [1]

Let C and D be model categories and $F: C \to D$ be a Reedy admissible functor. Then $C_F D$ admits the following model structure. A map $(A, B, FA \to B) \to (A', B', FA' \to B')$ is

— A weak equivalence if $A \to A'$ is a weak equivalence in C and $B \to B'$ is a weak equivalence in D,
— A fibration if $A \to A'$ is a fibration in C and $B \to B'$ is a fibration in D.
— A cofibration if $A \to A'$ is a cofibration in C and $FA' \cup_{FA} B \to B'$ is a cofibration in D.

Along the lines of Lemma 4.5 in [1] we establish the next result.

Lemma 4.5

The functor $\nabla : S^{\mathbb{Z}/2} \to S^d$ is Reedy admissible.

Hence we can define our model of O(2)-spaces as the category $\mathbf{S}^{\mathbb{Z}/2}{}_{\nabla}\mathbf{S}^d$ with the above model structure.

EQUIVALENCE OF HOMOTOPY CATEGORIES

The aim of the final section is to show that the category $S^{\mathbb{Z}/2}{}_{\nabla}S^d$ is indeed a model for $\text{Top}^{0(2)}$ with the following model structure:

Proposition 5.1

The category $\text{Top}^{0(2)}$ has a model structure such that a map $f : X \to Y$ is

— a weak equivalence if $f^H : X^H \to Y^H$ is a weak equivalence of spaces for all closed subgroups H in O(2),
— a fibration if $f^H : X^H \to Y^H$ is a Serre fibration for all closed subgroups H in O(2),
— a cofibration if it has the left lifting property with respect to acyclic fibrations.

Again, this follows from Theorem 2.2 in [11].

Theorem 5.2

There is a pair of adjoint functors

$$L : \mathbf{S}^{\mathbb{Z}/2}{}_{\nabla}\mathbf{S}^d \leftrightarrow \mathbf{Top}^{0(2)} : R.$$

The left adjoint $L: S^{\mathbb{Z}/2}_{\nabla} S^d \to \mathrm{top}^{0(2)}$ is defined as follows: given a triple $A = (A_{\mathbb{Z}/2}, A_d, \nabla A_{\mathbb{Z}/2} \to A_d)$, its image under L is the pushout of the diagram

$$|A_{\mathbb{Z}/2}|_{\mathbb{Z}/2} \xleftarrow{\zeta} |\nabla(A_{\mathbb{Z}/2})|_d \to |A_d|_d$$

where the map ζ is the map from Lemma 4.2.

Define the right adjoint $R: \mathrm{Top}^{0(2)} \to S^{\mathbb{Z}/2}_{\nabla} S^d$ as follows: given an O(2)-space X, its image under R is the triple $\left(S^*_{\mathbb{Z}/2}\left(X^{S^1}\right), S^*_d(X), \nabla\left(S^*_{\mathbb{Z}/2}\left(X^{S^1}\right)\right) \to S^*_d(X) \right)$, where the last element in the triple is the map adjoint to

$$|\nabla(S^*_{\mathbb{Z}/2}(X^{S^1}))|_d \to |S^*_{\mathbb{Z}/2}(X^{S^1})|_{\mathbb{Z}/2} \to X^{S^1} \to X$$

under the adjunction between the functors $|?|_{db}$ and S^*_d.

This is established using arguments similar to those in Section 5 of [1].

Theorem 5.3

The adjoints L and R satisfy the assumptions of part (1) of Quillen's equivalence theorem, and hence induce a pair of adjoints on the appropriate homotopy categories.

Proof

If $f: X \to Y$ is a fibration in $\mathrm{Top}^{0(2)}$, then $f^H: X^H \to Y^H$ is a fibration for all closed subgroups $H \subseteq O(2)$. Hence, applying Theorem 4.3 in [8], we conclude that $S^*_d(f): S^*_d(X) \to S^*_d(Y)$ is a fibration of dihedral sets. Moreover, $f^{S^1}: X^{S^1} \to Y^{S^1}$ is a fibration of $\mathbb{Z}/2$-spaces. Hence $S^*_{\mathbb{Z}/2}(f): S^*_{\mathbb{Z}/2}(X) \to S^*_{\mathbb{Z}/2}(Y)$

is a fibration of $\Delta(\mathbb{Z}/2)$-sets in the model structure of Proposition 3.4. Since fibrations in $S^{\mathbb{Z}/2}_{\nabla}S^d$ are defined coordinatewise, it follows that R(f) is a fibration. The proof that R preserves weak equivalences is similar.

We need the following analogue of Blumberg's Lemma 6.3:

Lemma 5.4

If f:X→Y is a cofibration of dihedral sets, then the maps

$$|f|^{O(2)}_d : |X|^{O(2)}_d \rightarrow |Y|^{O(2)}_d, \qquad |f|^{S^1}_d : |X|^{S^1}_d \rightarrow |Y|^{S^1}_d$$

are homeomorphisms.

Corollary 5.5

If X is a cofibrant dihedral set, then $|X|^{O(2)}_d$ and $|X|^{S^1}_d$ are empty.

Corollary 5.6

If $A=(A_{\mathbb{Z}/2},A_d,\nabla A_{\mathbb{Z}/2}\rightarrow A_d)$ is a cofibrant object in $S^{\mathbb{Z}/2}_{\nabla}S^d$, then $L(A)^{S^1}\simeq A_{\mathbb{Z}/2}$ and for all finite subgroups $H\subseteq O(2)$ we have $L(A)^H=|Ad|^H_d$.

The main result of this paper is the following:

Theorem 5.7

The adjoints L and R satisfy the assumptions of part (2) of Quillen's equivalence theorem and hence induce an equivalence of homotopy theories.

Proof

Let $A=(A_{\mathbb{Z}/2}, A_d, \nabla A_{\mathbb{Z}/2} \to A_d)$ be a cofibrant object in $S^{\mathbb{Z}/2} \underset{\nabla}{} S^d$ and let X be a fibrant object in $\text{Top}^{O(2)}$. We need to show that a map $f:A \to R(X)$ is a weak equivalence in $S^{\mathbb{Z}/2} \underset{\nabla}{} S^d$ if and only if its adjoint $f^\flat : L(A) \to X$ is a weak equivalence in $\text{Top}^{O(2)}$.

Let $f:A \to R(X)$ be a weak equivalence. Hence the maps $A_{\mathbb{Z}/2} \to S^*_{\mathbb{Z}/2}(X)$ and $A_d \to S^*_d(X)$ are weak equivalences of $\Delta(\mathbb{Z}/2)$-sets and dihedral sets respectively. Therefore their adjoints $|A_{\mathbb{Z}/2}|_{\mathbb{Z}/2} \to X$ and $|A_d|_d \to X$ are also weak equivalences. Applying the two out of three property and the previous corollary to the composite $|A_{\mathbb{Z}/2}|_{\mathbb{Z}/2} \to L(A) \to X$, we see that the map $f^\flat : L(A) \to X$ is a $\mathbb{Z}/2$-equivalence. Applying the same reasoning to the composite $|A_d|_d \to L(A) \to X$, we see that the map $f^\flat : L(A) \to X$ is an equivalence of $O(2)$-spaces with the model structure taking into account fixed point data of finite subgroups. We conclude that $f^\flat : L(A) \to X$ is a weak equivalence in $\text{Top}^{O(2)}$.

Now suppose that the map $f^\flat : L(A) \to X$ is a weak equivalence in $\text{Top}^{O(2)}$. By the previous corollary we know that the map $|A_{\mathbb{Z}/2}|_{\mathbb{Z}/2} \to L(A)$ is a $\mathbb{Z}/2$-equivalence. Hence the composite $|A_{\mathbb{Z}/2}|_{\mathbb{Z}/2} \to L(A) \to X$ is also a $\mathbb{Z}/2$-equivalence. By Section 3 we conclude that the map $A_{\mathbb{Z}/2} \to S^*_{\mathbb{Z}/2}(X)$ is a weak equivalence of $\Delta(\mathbb{Z}/2)$-sets. Again by the previous corollary we know that the map $|A_d|_d^H \to L(A)^H$ is a weak equivalence for all finite subgroups H in $O(2)$. Hence the composite $|A_d|_d^H \to L(A)^H \to X^H$ is also a weak equivalence. We conclude that the map $|A_d|_d \to X$ is a weak equivalence of topological spaces in the model structure taking fixed point data of finite subgroups into account. By the proof of Theorem 4.3 in [8] we conclude that the adjoint $A_d \to S^*_d(X)$ is also a weak equivalence. It follows that the map $f:A=(A_{\mathbb{Z}/2}, A_d, \nabla A_{\mathbb{Z}/2} \to A_d) \to R(X)$ is a weak equivalence in $S^{\mathbb{Z}/2} \underset{\nabla}{} S^d$.

ACKNOWLEDGMENTS

The author thanks the referee for a number of useful suggestions. He also thanks M. Intermont for helpful conversations on this subject and the Department of Mathematics of Kalamazoo College for its hospitality.

REFERENCES

1. A. Blumberg, A discrete model for S 1 -homotopy theory, J. Pure Appl. Algebra 210 (2007) 29–41.
2. A. Connes, Cohomologie cyclique et foncteurs Extn , C. R. Acad. Sci. Paris Sér. I Math. 296 (1983) 953–958.
3. W.G. Dwyer, M. Hopkins, D.M. Kan, Homotopy theory of cyclic sets, Trans. Amer. Math. Soc. 291 (1985) 281–289.
4. M. Bökstedt, W.C. Hsiang, I. Madsen, The cyclotomic trace and algebraic K-theory of spaces, Invent. Math. 111 (1993) 465–539.
5. J. Spaliński, Strong homotopy theory of cyclic sets, J. Pure Appl. Algebra 99 (1995) 35–52.
6. J.-L. Loday, Homologies diédrale et quaternionique, Adv. Math. 66 (1987) 119–148.
7. Z. Fiedorowicz, J.-L. Loday, Crossed simplicial groups and their associated homology, Trans. Amer. Math. Soc. 326 (1991) 57–87.
8. J. Spaliński, Homotopy theory of dihedral and quaternionic sets, Topology 39 (2000) 557–572.
9. D.G. Quillen, Homotopical Algebra, in: Lect. Notes in Math., vol. 43, Springer, 1967.
10. W.G. Dwyer, J. Spaliński, Homotopy theories and model categories, in: Handbook of Algebraic Topology, North Holland, Netherlands, 1995, pp. 73–126.
11. W.G. Dwyer, D.M. Kan, Singular functors and realization functors, Indag. Math. 46 (1984) 147–153.
12. J. Peter May, Simplicial Objects in Algebraic Topology, Van Nostrand, 1967.

CITATION

Jan Spaliński, A discrete model of -homotopy theory, Journal of Pure and Applied Algebra, Volume 214, Issue 1, January 2010, Pages 1-5, ISSN 0022-4049, http://dx.doi.org/10.1016/j.jpaa.2009.04.013.

Three Models for the Homotopy Theory of Homotopy Theories

Julia E. Bergner
Department of Mathematics, University of Notre Dame, Notre Dame, IN 46556, United States

2

ABSTRACT

Given any model category, or more generally any category with weak equivalences, its simplicial localization is a simplicial category which can rightfully be called the "homotopy theory" of the model category. There is a model category structure on the category of simplicial categories, so taking its simplicial localization yields a "homotopy theory of homotopy theories". In this paper we show that there are two different categories of diagrams of simplicial sets, each equipped with an appropriate definition of weak equivalence, such that the resulting homotopy theories are each equivalent to the homotopy theory arising from the model category structure on simplicial categories. Thus, any of these three categories with the respective weak equivalences could be considered a model for the homotopy theory of homotopy theories. One of them in particular, Rezk's complete Segal space model category structure on the category of simplicial spaces, is much more convenient from the perspective of making calculations and therefore obtaining information about a given homotopy theory.

INTRODUCTION

Classical homotopy theory considers topological spaces, up to weak homotopy equivalence. Eventually, the structure of the category of topological spaces making it possible to talk about its "homotopy theory" was axiomatized; it is known as a model category structure. In particular, given a model category structure on an arbitrary category, we can talk about its homotopy category. More generally, we can think about the "homotopy theory" given by that category with its particular class of weak equivalences, where the homotopy theory encompasses the homotopy category as well as higher-order information. One might ask what specifically is meant by a homotopy theory.

One answer to this question uses simplicial categories, which in this paper we will always take to mean categories enriched over simplicial sets. Given a model category M, taking its simplicial localization with respect to its subcategory of weak equivalences yields a simplicial category LM [9, 4.1]. The simplicial localization encodes the known homotopy-theoretic information of the model category, so one point of view is that this simplicial category is the homotopy theory associated to the model category structure. Set-theoretic issues aside, we can also construct the simplicial localization for any category with a subcategory of weak equivalences, so therefore we can speak of an associated homotopy theory even in this more general situation.

Given two homotopy theories, one can ask whether they are equivalent to one another in some natural sense. There is a notion of weak equivalence between two simplicial categories which is a simplicial analogue of an equivalence between categories. These weak equivalences are known as DK-equivalences, where the "DK" refers to the fact that they were first defined by Dwyer and Kan in [8]. In fact, there is a model category structure SC on the category of all (small) simplicial categories in which the weak equivalences are these DK-equivalences [3, 1.1]. The associated homotopy theory of simplicial categories is what we will refer to as the homotopy theory of homotopy theories.

In [17], Rezk takes steps toward finding a model other than that of simplicial categories for the homotopy theory of homotopy theories. He defines complete Segal spaces, which are simplicial spaces satisfying some nice properties (Definition 3.4 and Definition 3.6 below) and constructs a functor which assigns a complete Segal space to any simplicial category. He considers a model category structure CSS on the category of all simplicial spaces in which the weak equivalences are levelwise weak equivalences of simplicial sets and then localizes it in such a way that the local objects are the complete Segal spaces (Theorem 3.8).

However, Rezk does not construct a functor from the category of complete Segal spaces to the category of simplicial categories, nor does he discuss the model category SC. In this paper, we complete his work by showing that SC and CSS have equivalent homotopy theories. This result is helpful in that the weak equivalences between complete Segal spaces are easy to identify (see Proposition 3.11 below), unlike the weak equivalences between simplicial categories, and therefore making any kind of calculations would be much easier in CSS. Using terminology of Dugger [7], this model category CSS is a presentation for the homotopy theory of homotopy theories, since it is a localization of a category of diagrams of spaces.

In order to prove this result, we make use of an intermediate category. Consider the full subcategorySeCat of the category of simplicial spaces whose objects are simplicial spaces with a discrete simplicial set in degree zero. We will prove the existence of two model category structures on SeCat, each with the same class of weak equivalences. The first of these structures, which we denote SeCat$_c$, has as cofibrations the maps which are level wise cofibrations of simplicial sets. (An alternate proof of the existence of this model category structure is given by Hirschowitz and Simpson [13, 2.3]. They actually prove the existence of such a model category structure for Segal n-categories, whereas we consider only the case where n=1.) The second model category structure, which we denote SeCat$_f$, has as fibrations maps which can be thought of as localizations of levelwise fibrations of simplicial sets, although strictly speaking they cannot be obtained in this way. We use these model category structures to produce a chain of Quillen equivalences

$$SC \leftrightarrows SeCa_{tf} \rightleftarrows SeCa_{tc} \rightleftarrows CSS.$$

(In each case, the topmost arrow is the left adjoint of the adjoint pair.) Notice that we can obtain a single Quillen equivalence $SeCa_{tf} \rightleftarrows CSS$ via composition. Since Quillen equivalent model categories have DK-equivalent simplicial localizations (Proposition 2.8), all three of these categories with their respective weak equivalences give models for the homotopy theory of homotopy theories.

Organization of the Paper

We begin in Section 2 by recalling standard information about model category structures and simplicial objects. In Section 3, we state the definitions of simplicial categories, complete Segal spaces, and Segal categories, and we give some basic results about each. In Section 4, we set up some constructions on Segal precategories that we will need in order to prove our model category structures. In Section 5, we prove the existence of a model category structure $SeCa_{tc}$ on the category of Segal precategories which we then in Section 6 show is Quillen equivalent to Rezk's complete Segal space model category structure CSS. In Section 7, we prove the existence of the model category structure $SeCa_{tf}$ on the category of Segal precategories and prove that it is Quillen equivalent to $SeCa_{tc}$. We then show in Section 8 that $SeCa_{tf}$ is Quillen equivalent to the model category structure SC on simplicial categories. Section 9 contains the proofs of some technical lemmas.

BACKGROUND ON MODEL CATEGORIES AND SIMPLICIAL OBJECTS

Model Categories

Recall that a model category structure on a category C is a choice of three distinguished classes of morphisms: fibrations ($\twoheadrightarrow$), cofibrations ($\hookrightarrow$), and weak equivalences ($\xrightarrow{\sim}$). A (co) fibration which is also a weak equivalence is an acyclic (co)fibration . With this choice of three classes of morphisms, C is required to satisfy five axioms MC1–MC5 which can be found in [10, 3.3].

In all the model categories we use, the factorizations given by axiom MC5 can be chosen to be functorial [14, 1.1.1]. An object X in a model category is fibrant if the unique map X→∗ to the terminal object is a fibration. Dually, X is cofibrant if the unique map from the initial object φ→X is a cofibration. Given any object X, the functorial factorization of the map X→∗ as the composite of an acyclic cofibration followed by a fibration

$$X \overset{\sim}{\hookrightarrow} X^f \twoheadrightarrow *$$

gives us the object X^f, the fibrant replacement of X. Dually, we can define its cofibrant replacement X^c using the functorial factorization

$$\phi \hookrightarrow X^c \overset{\sim}{\twoheadrightarrow} X.$$

All the model category structures that we work with are cofibrantly generated. In a cofibrantly generated model category, there are two sets of specified morphisms, the generating cofibrations and the generating acyclic cofibrations, such that a map is an acyclic fibration if and only if it has the right lifting property with respect to the generating cofibrations, and a map is a fibration if and only if it has the right lifting property with respect to the generating acyclic cofibrations [12, 11.1.2]. To prove that a particular category with a choice of weak equivalences has a cofibrantly generated model category structure, we need the following definition.

Definition 2.2: [12, 10.5.2]. Let C be a category and I a set of maps in C. Then an I-injective is a map which was the right lifting property with respect to every map in I. An I-cofibration is a map with the left lifting property with respect to every I-injective.

We are now able to state the theorem that we use in this paper to prove the existence of specific model category structures.

Theorem 2.3: [12, 11.3.1]. Let M be a category with a specified class of weak equivalences which satisfies model category axioms MC1 and MC2. Suppose further that the class of weak equivalences is closed under retracts. Let I and J be sets of maps in M which satisfy the following conditions:

— Both I and J permit the small object argument [12, 10.5.15].
— Every J-cofibration is an I-cofibration and a weak equivalence.
— Every I-injective is a J-injective and a weak equivalence.
— One of the following conditions holds:
  • A map that is an I-cofibration and a weak equivalence is a J-cofibration, or
  • A map that is both a J-injective and a weak equivalence is an I-injective.

Then there is a cofibrantly generated model category structure on Min which I is a set of generating cofibrations and J is a set of generating acyclic cofibrations.

We now define our notion of "equivalence" between two model categories. Recall that for categories C and D a pair of functors

$$F : \mathcal{C} \rightleftarrows \mathcal{D} : R$$

is an adjoint pair if for each object X of C and object Y of D there is an isomorphism $\varphi : \mathrm{Hom}_D(FX, Y) \to \mathrm{Hom}_C(X, RY)$ which is natural in X and Y [15, IV.1].

Definition 2.4: [14, 1.3.1]

If C and D are model categories, then the adjoint pair

$$F : \mathcal{C} \rightleftarrows \mathcal{D} : R$$

is a Quillen pair if one of the following equivalent statements is true:

— F preserves cofibrations and acyclic cofibrations.
— R preserves fibrations and acyclic fibrations.

Definition 2.5: [14, 1.3.12]: A Quillen pair is a Quillen equivalence if for all cofibrant X in C and fibrant Y in D, a map f: FX→Y is a weak equivalence in D if and only if the map φf:X→RY is a weak equivalence in C.

We will use the following proposition to prove that a Quillen pair is a Quillen equivalence. Recall that a functor F: C→Dreflects a property if, for any morphism f of C, whenever Ff has the property, then so does f.

Proposition 2.6: [14, 1.3.16].

Suppose that

$$F : \mathcal{C} \rightleftarrows \mathcal{D} : R$$

is a Quillen pair. Then the following statements are equivalent:

— This Quillen pair is a Quillen equivalence.
— Freflects weak equivalences between cofibrant objects and, for every fibrant Y in D, the map F ((RY) c) →Y is a weak equivalence.
— Rreflects weak equivalences between fibrant objects and, for every cofibrant X in C, the map X→R ((FX)f) is a weak equivalence.

The existence of a Quillen equivalence between two model categories is actually a stronger condition than we need, but it is a convenient way to show that two homotopy theories are the same. Here, we take the viewpoint that simplicial categories are models for homotopy theories. A simplicial category is a category C enriched over simplicial sets, or a category such that, for objects x and y of C, there is a simplicial set of morphisms Hom$_C$(x,y) between them. We will use the following notion of equivalence of simplicial categories.

Definition 2.7 [8, 2.4]

A functor f: C→D between two simplicial categories is a DK-equivalence if it satisfies the following two conditions:

— for any objects x and y of C, the induced map $\mathrm{Hom}_C(x,y) \to$ $\mathrm{Hom}_D(fx,fy)$ is a weak equivalence of simplicial sets, and
— The induced map of categories of components $\varpi_0 f : \varpi_0 C \to \varpi 0 D$ is an equivalence of categories.

Recall that the category of components $\pi_0 C$ of a simplicial category C is the category with the same objects as C and such that

$$\mathrm{Hom}_{\pi_0 \mathcal{C}}(x, y) = \pi_0 \mathrm{Hom}_{\mathcal{C}}(x, y).$$

Now, the following result tells us that model categories which are Quillen equivalent actually have equivalent homotopy theories.

Proposition 2.8 [8, 5.4]: Suppose that C and D are Quillen equivalent model categories. Then the simplicial localizations LC and LD are DK-equivalent.

Simplicial Objects

Recall that a simplicial set is a functor $\Delta^{\mathrm{op}} \to$ Sets, where the cosimplicial category Δ has as objects the finite ordered sets [n]={0,...,n} and as morphisms the order-preserving maps, and Δ^{op} is its opposite category. In particular, for n≥0, we have Δ[n], the n-simplex, $\Delta\big[n\big]$, the boundary of Δ[n], and, for n>0 and 0≤k≤n, V[n,k], which is $\Delta\big[n\big]$ with the kth face removed [11, I.1]. For any simplicial set X, we denote by X_n the image of [n]. There are face maps $d_i : X_n \to X_n-1$ for 0≤i≤n and degeneracy maps $s_i : X_n \to X_n+1$ for 0≤i≤n, satisfying certain compatibility conditions [11,I.1]. We denote by |X| the topological space given by geometric realization of the simplicial set X [11, I.2].

There is a model category structure on simplicial sets in which the weak equivalences are the maps which become weak homotopy equivalences of topological spaces after geometric realization [11, I.11.3]. We denote this model category structure by SSets. Note in particular that it is cofibrantly generated. The generating cofibrations are the maps $\Delta[m] \to \Delta[m]$ for all $m \geq 0$, and the generating acyclic cofibrations are the maps $V[m,k] \to \Delta[m]$ for all $m \geq 1$ and $0 \leq k \leq m$ [14, 3.2.1]. This model category structure is Quillen equivalent to the standard model category structure on topological spaces [14, 3.6.7]. In light of this fact, we will sometimes refer to simplicial sets as "spaces".

More generally, a simplicial object in a category C is a functor $\Delta^{\mathrm{op}} \to C$ [14, 3.1]. In particular, a simplicial space (or bisimplicial set) is a functor $\Delta^{\mathrm{op}} \to SSets$ [11, IV.1]. Given a simplicial set X, we also use X to denote the constant simplicial space with the simplicial set X in each degree. By X^t we denote the simplicial space such that $(X^t)n$ is the constant simplicial set X_n, or the simplicial set which has the set X_n in each degree.

Notice, however, that our definition of "simplicial category" in this paper is inconsistent with this terminology. There is a more general notion of simplicial category by which is meant a simplicial object in the category of small categories. Such a simplicial category is a functor $\Delta^{\mathrm{op}} \to Cat$ where Cat is the category with objects the small categories and morphisms the functors between them. Our definition of simplicial category coincides with this one when the extra condition is imposed that the face and degeneracy maps be the identity map on objects [8, 2.1].

We also require the following additional structure on some of our model category structures. A simplicial model category is a model category which is also a simplicial category satisfying two additional axioms [12, 9.1.6]. (Again, the terminology is potentially confusing because a simplicial model category is not a simplicial object in the category of model categories.) The important part of this structure that we use is the fact that, given objects X and Y of a simplicial model category, it makes sense to talk about the function complex , or simplicial set Map(X,Y).

Given a model category M, or more generally a category with weak equivalences, a homotopy function complex $Map^h(X,Y)$ is a simplicial set which is the morphism space between X and Y in the simplicial localization LM [8, Section 4]. If M is a simplicial model category, X is cofibrant in M, and Y is fibrant in M, then Map^h(X,Y) is weakly equivalent to Map(X,Y).

Localized Model Category Structures

Several of the model category structures that we use are obtained by localizing a given model category structure with respect to a map or a set of maps. Suppose that S={f: A→B} is a set of maps with respect to which we would like to localize a model category (or category with weak equivalences) M. We define an S-local object W to be an object of M such that for any f: A→B in S, the induced map on homotopy function complexes

$$f^*: \mathrm{Map}^h(B, W) \to \mathrm{Map}^h(A, W)$$

is a weak equivalence of simplicial sets. (If M is a model category, a local object is usually required to be fibrant.) A map g: X→Y in M is then defined to be an S-local equivalence if for every local object W, the induced map on homotopy function complexes

$$g^*: \mathrm{Map}^h(Y, W) \to \mathrm{Map}^h(X, W)$$

is a weak equivalence of simplicial sets.

The following theorem holds for model categories M which are left proper and cellular. We will not define these conditions here, but refer the reader to [12, 13.1.1, 12.1.1] for more details. We do note, in particular, that a cellular model category is cofibrantly generated. All the model categories that we localize in this paper can be shown to satisfy both these conditions.

Theorem 2.11: [12, 4.1.1].

Let M be a left proper cellular model category. There is a model category structure L$_S$M on the underlying category of M such that:

— The weak equivalences are the S-local equivalences.
— The cofibrations are precisely the cofibrations of M.
— The fibrations are the maps which have the right lifting property with respect to the maps which are both cofibrations and S-local equivalences.
— The fibrant objects are the S-local objects which are fibrant in M.
— If M is a simplicial model category, then its simplicial structure induces a simplicial structure on $_{LS}$M.

In particular, given an object X of M, we can talk about its functorial fibrant replacement LX in $_{LS}$M. The object LX is an S-local object which is fibrant in M, and we will refer to it as the localization of X in $_{LS}$M.

Model Category Structures for Diagrams of Spaces

Suppose that D is a small category and consider the category of functors D→SSets, denoted SSets^D. This category is also called the category of D-diagrams of spaces. We would like to consider model category structures on SSets^D.

A natural choice for the weak equivalences in SSets^D is the class of levelwise weak equivalences of simplicial sets. Namely, given two D-diagrams X and Y, we define a map f: X→Y to be a weak equivalence if and only if for each object d of D, the map X(d)→Y(d) is a weak equivalence of simplicial sets.

There is a model category structure SSets^D_f on the category of D-diagrams with these weak equivalences and in which the fibrations are given by levelwise fibrations of simplicial sets. The cofibrations in SSets^D_f are then the maps of simplicial spaces which have the left lifting property with respect to the maps which are levelwise acyclic fibrations. This model structure is often called the projective model category structure on D-diagrams of spaces [11, IX, 1.4]. Dually, there is a model category structure SSets^D_c in which the cofibrations are given by level wise cofibrations of simplicial sets, and this model structure is often called the injective model category structure [11, VIII, 2.4]. The small category D which we use in this paper is Δ^{op}, so that the diagram category $SSet^{\Delta^{op}}$ is just the category of simplicial spaces.

Consider the Reedy model category structure on simplicial spaces [16]. In this structure, the weak equivalences are again the level wise weak equivalences of simplicial sets. The Reedy model category structure is cofibrantly generated, where the generating cofibrations are the maps

$$\dot{\Delta}[m] \times \Delta[n]^t \cup \Delta[m] \times \dot{\Delta}[n]^t \to \Delta[m] \times \Delta[n]^t$$

for all n,m≥0. The generating acyclic cofibrations are the maps

$$V[m, k] \times \Delta[n]^t \cup \Delta[m] \times \dot{\Delta}[n]^t \to \Delta[m] \times \Delta[n]^t$$

for all n≥0, m≥1, and 0≤k≤m [17, 2.4].

It turns out that the Reedy model category structure on simplicial spaces is exactly the same as the injective model category structure on this same category, as given by the following result.

Proposition 2.14 [12, 15.8.7, 15.8.8]: A map f: X→Yof simplicial spaces is a cofibration in the Reedy model category structure if and only if it is a monomorphism. In particular, every simplicial space is Reedy cofibrant.

In light of this result, we denote the Reedy model structure on simplicial spaces by $SSet_C^{\Delta^{op}}$. Both $SSet_C^{\Delta^{op}}$ and $SSet_f^{\Delta^{op}}$ are simplicial model categories. In each case, given two simplicial spaces X and Y, we can define Map(X,Y) by

$$Map(X, Y)_n = Hom(X \times \Delta[n], Y)$$

where the set on the right-hand side consists of maps of simplicial spaces.

To establish some notation we need later in the paper, we recall the definition of fibration in the Reedy model category structure. If X is a simplicial space, let $sk_n X$ denote its n-skeleton, generated by the spaces in degrees less than or equal to n, and let $cosk_n X$ denote the n-coskeleton of X [16, Section 1]. A map X→Y is a fibration in $SSet_c^{\Delta^{op}}$ if

— $X_0 \to Y_0$ is a fibration of simplicial sets, and
— for all n≥1, the map $X_n \to P_n$ is a fibration, where P_n is defined to be the pullback in the following diagram:

$$
\begin{array}{ccc}
P_n & \longrightarrow & Y_n \\
\downarrow & & \downarrow \\
(cosk_{n-1} X)_n & \longrightarrow & (cosk_{n-1} Y)_n.
\end{array}
$$

Notice in particular that this pullback diagram is actually a homotopy pullback diagram, as follows. If f: X→Y is a Reedy fibration, then it has the right lifting property with respect to all Reedy acyclic cofibrations. In particular, there is a dotted arrow lift in the following diagram, where m≥1, 0≤k≤m, and n≥0:

$$
\begin{array}{ccc}
V[m,k] \times \dot\Delta[n]^t & \longrightarrow & X \\
\downarrow & \nearrow & \downarrow \\
\Delta[m] \times \dot\Delta[n]^t & \longrightarrow & Y.
\end{array}
$$

Since the functors sk_n and $cosk_n$ are adjoint [16, Section 1], we have that $(cosk_{n-1} X)_n \simeq Map(\Delta[n], cosk_n X) \simeq Map(sk_n \Delta[n], X) \simeq Map(\dot\Delta[n], X)$.

Therefore, we have a dotted arrow lift in each diagram

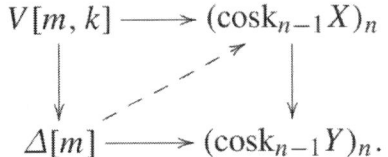

$$V[m, k] \longrightarrow (\mathrm{cosk}_{n-1}X)_n$$

In particular, the right-hand vertical arrow is a fibration of simplicial sets. Thus, the simplicial set P_n is a homotopy pullback and therefore homotopy invariant.

We also make use of the projective model category structure $\mathrm{SSets}_f^{\Delta^{op}}$ on simplicial spaces. This model category is also cofibrantly generated; the generating cofibrations are the maps

$$\dot{\Delta}[m] \times \Delta[n]^t \to \Delta[m] \times \Delta[n]^t$$

for all m,n≥0 [11, IV.3.1].

In the next section, we localize the Reedy (or injective) and projective model category structures on simplicial spaces with respect to a map to obtain model category structures in which the fibrant objects are Segal spaces (Definition 3.4). We will further localize them to obtain model category structures in which the fibrant objects are complete Segal spaces (Definition 3.6).

SOME DEFINITIONS AND MODEL CATEGORY STRUCTURES

In this section, we define and discuss in turn the three main structures that we will use in the course of this paper: simplicial categories, complete Segal spaces, and Segal categories.

Simplicial Categories

Simplicial categories, most simply stated, are categories enriched over simplicial sets, or categories with a simplicial set of morphisms between any two objects. So, given any objects x and y in a simplicial category C, there is a simplicial set $\text{Hom}_C(x,y)$.

Fix an object set O and consider the category of simplicial categories with object set O such that all morphisms are the identity on the objects. Dwyer and Kan define a model category structure SC_O in which the weak equivalences are the functors f: C→D of simplicial categories such that given any objects x and y of C, the induced map

$$\text{Hom}_C(x, y) \to \text{Hom}_D(x, y)$$

is a weak equivalence of simplicial sets [9, Section 7]. The fibrations are the functors f: C→D for which these same induced maps are fibrations, and the cofibrations are the functors which have the left lifting property with respect to the acyclic fibrations.

It is more useful, however, to consider the category of all small simplicial categories with no restriction on the objects. Before describing the model category structure on this category, we need a few definitions. Recall from Definition 2.7 above that if C is a simplicial category, then we denote by $\pi_0 C$ the category of components of C.

If C is a simplicial category and x and y are objects of C, a morphism $e \in \text{Hom}_C(x,y)_0$ is a homotopy equivalence if the image of e in $\pi_0 C$ is an isomorphism.

Theorem 3.2 [3, 1.1]: There is a model category structure on the category SC of small simplicial categories defined by the following three classes of morphisms:

— The weak equivalences are the maps f: C→D satisfying the following two conditions:
  • (W_1) For any objects x and y in C, the map

$$\text{Hom}_{\mathcal{C}}(x, y) \rightarrow \text{Hom}_{\mathcal{D}}(fx, fy)$$

is a weak equivalence of simplicial sets.

- (W$_2$) The induced functor $\varpi_0 f : \varpi_0 C \rightarrow \varpi_0 D$ on the categories of components is an equivalence of categories.

— The fibrations are the maps f: C→D satisfying the following two conditions:

- (F$_1$) For any objects x and y in C, the map

$$\text{Hom}_{\mathcal{C}}(x, y) \rightarrow \text{Hom}_{\mathcal{D}}(fx, fy)$$

is a fibration of simplicial sets.

- (F$_2$) For any object x_1 in C, y in D, and homotopy equivalence e: $fx_1 \rightarrow y$ in D, there is an object x_2 in C and homotopy equivalence d: $x_1 \rightarrow x_2$ in C such that fd=e.

— The cofibrations are the maps which have the left lifting property with respect to the maps which are fibrations and weak equivalences.

Notice that the weak equivalences are precisely the DK-equivalences that we defined above (Definition 2.7).

The proof of this theorem actually shows that this model category structure is cofibrantly generated. Define the functor U: SSets→SC such that for any simplicial set K, the simplicial category UK has two objects, x and y, and only nonidentity morphisms the simplicial set K=Hom (x,y). Using this functor, we define the generating cofibrations to be the maps of simplicial categories

- (C$_1$) $U \cdot \dot\Delta[n] \rightarrow U\Delta[n]$ for n≥0, and
- (C$_2$) $\varphi \rightarrow \{x\}$, where φ is the simplicial category with no objects and $\{x\}$ denotes the simplicial category with one object x and no nonidentity morphisms.

The generating acyclic cofibrations are defined similarly [3, Section 1].

Segal Spaces and Complete Segal Spaces

Complete Segal spaces, defined by Rezk in [17], are more difficult to describe, but ultimately they are actually easier to work with than simplicial categories. The name "Segal" refers to the similarity between Segal spaces and Segal's Γ-spaces [18].

We begin by defining Segal spaces. In [17, 4.1], Rezk defines for each $0 \leq i \leq k-1$ a map $\alpha^i \colon [1] \to [k]$ in Δ such that $0 \mapsto i$ and $1 \mapsto i+1$. Then for each k he defines the simplicial space

$$G(k)^t = \bigcup_{i=0}^{k-1} \alpha^i \Delta[1]^t \subset \Delta[k]^t.$$

He shows that, for any simplicial space X, there is an weak equivalence of simplicial sets

$$\mathrm{Map}^h_{\mathcal{SS}ets^{\Delta^{op}}}(G(k)^t, X) \to \underbrace{X_1 \times^h_{X_0} \cdots \times^h_{X_0} X_1}_{k},$$

where the right-hand side is the homotopy limit of the diagram

$$X_1 \xrightarrow{d_0} X_0 \xleftarrow{d_1} X_1 \xrightarrow{d_0} \cdots \xrightarrow{d_0} X_0 \xleftarrow{d_1} X_1$$

with k copies of X_1.

Now, given any k, define the map $\varphi^k \colon G(k)^t \to \Delta[k]^t$ to be the inclusion map. Then for any simplicial space W there is a map

$$\varphi_k = \mathrm{Map}^h_{\mathcal{SS}ets^{\Delta^{op}}}(\varphi^k, W) \colon \mathrm{Map}^h_{\mathcal{SS}ets^{\Delta^{op}}}(\Delta[k]^t, W) \to \mathrm{Map}^h_{\mathcal{SS}ets^{\Delta^{op}}}(G(k)^t, W).$$

More simply written, this map is

$$\varphi_k \colon W_k \to \underbrace{W_1 \times^h_{W_0} \cdots \times^h_{W_0} W_1}_{k}$$

and is often called a Segal map.

Definition 3.4 [17, 4.1]:
A Reedy fibrant simplicial space W is a Segal space if for each $k \geq 2$ the map φ^k is a weak equivalence of simplicial sets. In other words, the Segal maps

$$\varphi_k : W_k \to \underbrace{W_1 \times^h_{W_0} \cdots \times^h_{W_0} W_1}_{k}$$

are weak equivalences for all $k \geq 2$.

Notice that if W is a Segal space, or more generally if W is Reedy fibrant, we can use ordinary function complexes and a limit in the definition of the Segal maps [17, Section 4].

Rezk defines the coproduct of all these inclusion maps

$$\varphi = \coprod_{k \geq 0} (\varphi^k : G(k)^t \to \Delta[k]^t).$$

Using this map φ, we have the following result.

Theorem 3.5 [17, 7.1]:
There is a model category structure on simplicial spaces which can be obtained by localizing the Reedy model category structure with respect to the map φ. This model category structure has the following properties:

— The weak equivalences are the maps f for which $\mathrm{Map}^h_{\mathrm{SSets}^{\Delta^{op}}}(f, W)$ is a weak equivalence of simplicial sets for any Segal space W.
— The cofibrations are the monomorphisms.
— The fibrant objects are the Reedy fibrant φ-local objects, which are precisely the Segal spaces.

We will refer to this model category structure on simplicial spaces as the Segal space model category structure and denote it SeSpc.

The properties of Segal spaces enable us to speak of them much in the same way that we speak of categories. Heuristically, a simple example of a Segal space is the nerve of a category C, regarded as a simplicial space $nerve(C)^t$. (We need to take a Reedy fibrant replacement of this nerve to be an actual Segal space.) In particular, we can define "objects" and "maps" of a Segal space. We summarize the particular details here that we need; a full description is given by Rezk [17, Section 5].

Given a Segal space W, define its set of objects, denoted ob (w), to be the set of 0-simplices of the space W_0, namely, the set $W_{0,0}$. Given any two objects x,y in ob(w), define the mapping space map_w (x,y)to be the homotopy fiber of the map $(d_1, d_0): W_1 \to W_0 \times W_0$ over (x,y). (Note that since W is Reedy fibrant, this map is a fibration, and therefore in this case we can just take the fiber.) Given a 0-simplex x of W0, we denote by id_x the image of the degeneracy map $s_0: W_0 \to W_1$. We say that two 0-simplices of map_w(x,y), say f and g, are homotopic, denoted f~g, if they lie in the same component of the simplicial set map_w(x,y).

Given $f \in map_w(x, y)_0$ and $g \in map_w(y, z)_0$, there is a composite $g \circ f \in map_w(x, z)_0$, and this notion of composition is associative up to homotopy. We define the homotopy category $H_0(W)$ of W to have as objects the set ob(W) and as morphisms between any two objects x and y, the set $map_{H_0(W)}(x, y) = \pi_0 map_w(x, y)$.

A map g in $map_w(x,y)_0$ is a homotopy equivalence if there exist maps $f, h \in map_w(y, x)_0$ such that $g \circ f \sim id_y$ and $h \circ g \sim id_x$. Any map in the same component as a homotopy equivalence is itself a homotopy equivalence [17, 5.8]. Therefore we can define the space $W_{hoequiv}$ to be the subspace of W_1 given by the components whose zero-simplices are homotopy equivalences.

We then note that the degeneracy map $s_0: W_0 \to W_1$ factors through $W_{hoequiv}$ since for any object x the map $s_0(x) = id_x$ is a homotopy equivalence. Therefore, we have the following definition:

Definition 3.6 [17, Section 6]: A complete Segal space is a Segal space W for which the map $s_0 : W_0 \to W_{hoequiv}$ is a weak equivalence of simplicial sets.

We now consider an alternate way of defining a complete Segal space which is less intuitive but will enable us to localize the Segal space model category structure further in such a way that the complete Segal spaces are the new fibrant objects. Consider the category I[1] which consists of two objects x and y and exactly two non-identity maps which are inverse to one another, x→y and y→x. Denote by E the nerve of this category, and by Et the corresponding simplicial space. There are two maps $\Delta[0]^t \to E^t$ given by the inclusions of $\Delta[0]$t to the objects x and y, respectively. Let $\Psi: \Delta[0]^t \to E^t$ be the map which takes $\Delta[0]^t$ to the object x. (It does not actually matter which one of the two maps we have chosen, as long as it is fixed.) This map then induces, for any Segal space W, a map on homotopy function complexes

$$\psi^* : \mathrm{Map}^h_{\mathcal{SS}ets^{\Delta^{op}}}(E^t, W) \to \mathrm{Map}^h_{\mathcal{SS}ets^{\Delta^{op}}}(\Delta[0]^t, W) = W_0.$$

Proposition 3.7 [17, 6.4]: For any Segal space W, the map Ψ_* of homotopy function complexes is a weak equivalence of simplicial sets if and only if W is a complete Segal space.

Given this proposition, we can further localize the category of simplicial spaces with respect to this map.

Theorem 3.8 [17, 7.2]: Taking the localization of the Reedy model category structure on simplicial spaces with respect to the maps φ and Ψ above results in a model category structure which satisfies the following properties:

— The weak equivalences are the maps f such that $\mathrm{map}^h_{\mathrm{SSets}^{\Delta^{op}}}(f, W)$ is a weak equivalence of simplicial sets for any complete Segal space W.
— The cofibrations are the monomorphisms.
— The fibrant objects are the complete Segal spaces.

We refer to this model category structure on simplicial spaces as the complete Segal space model category structure, denoted CSS. It turns out that when the objects involved are Segal spaces, the weak equivalences in this model category structure can be described more explicitly.

Definition 3.9: A map $f: U \to V$ of Segal spaces is a DK-equivalence if

— for any pair of objects $x, y \in U_0$, the induced map $\mathrm{map}_U(x, y) \to \mathrm{map}_V(fx, fy)$ is a weak equivalence of simplicial sets, and
— the induced map $H_0(f): H_0(U) \to H_0(V)$ is an equivalence of categories.

We then have the following result by Rezk:

Theorem 3.10 [17, 7.7]: Let $f: U \to V$ be a map of Segal spaces. Then f is a DK-equivalence if and only if it becomes a weak equivalence in CSS.

Note that these weak equivalences have been given the same name as the ones in SC. While this may at first seem strange, the two definitions are very similar, in fact rely on the same generalization of the idea of equivalence of categories to a simplicial setting.

However, what is especially nice about the complete Segal space model category structure is the simple characterization of the weak equivalences between the fibrant objects.

Proposition 3.11 [17, 7.6]: A map $f: U \to V$ between complete Segal spaces is a DK-equivalence if and only if it is a level wise weak equivalence.

This proposition is actually a special case of a more general result. In any localized model category structure, a map is a local equivalence between fibrant objects if and only if it is a weak equivalence in the original model category structure [12, 3.2.18].

It is also possible to localize the projective model category structure $\mathrm{SSets}_f^{\Delta^{op}}$ on the category of simplicial spaces to obtain analogous model category structures. We will denote the localization of the projective model category structure by with respect to the map φ by SeS_{pf}. There is also a localization of the projective model category structure with respect to the maps φ and Ψ analogous to the model category structure CSS, but we do not need this structure here.

Segal Categories

Lastly, we consider the Segal categories. We begin by defining the preliminary notion of a Segal precategory.

Definition 3.13 [13, Section 2]: A Segal precategory is a simplicial space X such that the simplicial set X_0 in degree zero is discrete, i.e., a constant simplicial set.

In the case of Segal precategories, it again makes sense to talk about the Segal maps

$$\varphi_k : X_k \to \underbrace{X_1 \times_{X_0}^h \cdots \times_{X_0}^h X_1}_{k}$$

for each $k \geq 2$. Since X0 is discrete, we can actually take the limit

$$\underbrace{X_1 \times_{X_0} \cdots \times_{X_0} X_1}_{k}$$

on the right-hand side.

Definition 3.14 [13, Section 2]: A Segal category X is a Segal precategory such that each Segal map$_{\varphi k}$ is a weak equivalence of simplicial sets for $k \geq 2$.

Note that the definition of a Segal category is similar to that of a Segal space, with the additional requirement that the degree zero space be discrete. However, Segal categories are not required to be Reedy fibrant, so they are not necessarily Segal spaces.

Given a fixed set O, we can consider the category $SSets_O^{\Delta^{op}}$ whose objects are the Segal precategories with O in degree zero and whose morphisms are the identity on this set. There is a model category structure $SSets_{O,f}^{\Delta^{op}}$ on this category in which the weak equivalences are level wise [5, 3.7]. In other words, f: X→Y is a weak equivalence if for each n≥0, the map f_n: $X_n \to Y_n$ is a weak equivalence of simplicial sets. Furthermore, the fibrations are also level wise. This model structure can then be localized with respect to a map similar to the map which we used to obtain the Segal space model category structure.

We first need to determine what this map should be. We begin by considering the maps of simplicial spaces $\varphi k: G(k)^t \to \Delta[k]^t$ and adapting them to the case at hand.

The first problem is that $\Delta[k]t$ is not going to be in $SSets_{O,f}^{\Delta^{op}}$ for all values of k. Instead, we need to define a separate k-simplex for any k-tuple $x_0, \ldots, x_k$ of objects in O, denoted $\Delta[k]_{x_0,\ldots,x_k}^t$, so that the objects are preserved. Note that this object $\Delta[k]_{x_0,\ldots,x_k}^t$ also needs to have all elements of O as 0-simplices, so we add any of these elements that have not already been included in the x_i's, plus their degeneracies in higher degrees.

Then we can define

$$G(k)_{x_0,\ldots,x_k}^t = \bigcup_{i=0}^{k-1} \alpha^i \Delta[1]_{x_i,x_{i+1}}^t.$$

Now, we need to take coproducts not only over all values of k, but also over all k-tuples of vertices. Hence, the resulting map φ_O looks like

$$\varphi_O = \coprod_{k \geq 0} \left(\coprod_{(x_0,\ldots,x_k) \in O^{k+1}} (G(k)_{x_0,\ldots,x_k}^t \to \Delta[k]_{x_0,\ldots,x_k}^t) \right).$$

Setting $\underline{x} = (x_o......x_k)$, we can write the component maps as $G(k)^t_{\underline{x}} \to \Delta(k)^t_{\underline{x}}$. We can then localize $SSets_{O,f}^{\Delta^{op}}$ with respect to the map φ_O to obtain a model category which we denote $LSSets_{O,f}^{\Delta^{op}}$.

There are also analogous model category structures $SSets_{O,c}^{\Delta^{op}}$ and $LSSets_{O,c}^{\Delta^{op}}$ on the category of Segal precategories with a fixed set O in degree zero with the same weak equivalences but where the cofibrations, rather than the fibrations, are defined levelwise, and then we can localize with respect to the same map [5, 3.9], [19, A.1.1].

However, we would like a model category structure on the category of all Segal precategories, not just on these more restrictive subcategories. In the course of this paper, we prove the existence of two model category structures on Segal precategories. Unlike in the fixed object set case, we cannot actually obtain the model category structure via localization of a model category structure with level wise weak equivalences since it is not possible to put a model structure on the category of Segal precategories in which the weak equivalences are level wise and in which the cofibrations are monomorphisms.

To see that there is no such model structure, suppose that one did exist and consider the map $f: \Delta[0]^t\Delta[0]^t \to \Delta[0]^t$. By model category axiom MC5, f could be factored as the composite of a cofibration $\Delta[0]^t\Delta[0]^t \to X$ followed by an acyclic fibration $X \to \Delta[0]^t$. However, since the weak equivalences would be levelwise weak equivalences, X_0 would have to consist of one point. However, the only map $(\Delta[0]^t\Delta[0]^t)_0 \to X_0$ is not a monomorphism. Thus, there is no such factorization of the map f, and therefore there can be no model category structure satisfying the two given properties.

Relationship between Simplicial Categories and Segal Categories in Fixed Object Set Cases

Recall from above that there is a model category structure SCO on the category whose objects are the simplicial categories with a fixed set O of objects and whose morphisms are the functors which are the identity on the objects and that there is a model category structure

$LSSets_{O,c}^{\Delta^{op}}$ on the category whose objects are the Segal precategories with the set O in degree zero and whose morphisms are the identity on degree zero.

Theorem 3.16 [5, 5.5] There is an adjoint pair

$$F_O : LSSets_{O,f}^{\Delta^{op}} \rightleftarrows SC_O : R_O$$

which is a Quillen equivalence.

The proof of this theorem uses a generalization of a result by Badzioch [1, 6.5] which relates strict and homotopy algebras over an algebraic theory. This generalization uses the notion of multi-sorted algebraic theory [4].

A key step in this proof is a explicit description of the localization of the objects $\Delta[n]_{\underline{x}}^t$. Up to homotopy, this localization is the same as the localization of the objects $G[n]_{\underline{x}}^t$ and is obtained by taking the colimit of stages of a filtration

$$G(n)_{\underline{x}}^t = \Psi_1 G(n)_{\underline{x}}^t \subseteq \Psi_2 G(n)_{\underline{x}}^t \subseteq \cdots .$$

Let e_i denote the nondegenerate 1-simplex $x_i-1 \to x_i$ in $G(n)_{\underline{x}}^t$ and let w_i denote a word in the e_i's which can be obtained via "composition" of these 1-simplices. The k-th stage of the filtration is given by

$$(\Psi_k(G(n)_{\underline{x}}^t))_m = \{(w_1|\cdots|w_m) \mid \ell(w_1\cdots w_m) \leq k\}$$

where $\ell(w_1\cdots w_n)$ denotes the length of the word $w_1\cdots w_n$. The colimit of this filtration is weakly equivalent to $L_cG(n)_{\underline{x}}^t$ in $LSSets_{O,f}^{\Delta^{op}}$.

We show in the proof of [5, 4.2] that for each i≥1 the map

$$\Psi_i G(n)_{\underline{x}}^t \rightarrow \Psi_{i+1}G(n)_{\underline{x}}^t$$

is a DK-equivalence, and that the unique map from $G(n)_{\underline{x}}^t$ to the co-limit of this directed system is also a DK-equivalence.

In the current paper, we use some of the ideas of the proof from the fixed object set case, but we no longer use multi-sorted theories as we pass from SC_O to SC and $SSets_O^{\Delta^{op}}$ to $SeCat$.

METHODS OF OBTAINING SEGAL PRECATEGORIES FROM SIMPLICIAL SPACES

In the course of proving the existence of these two model category structures $SeCa_{tc}$ and $SeCa_{tf}$, we need sets of generating cofibrations and generating acyclic cofibrations which are similar to those of the Reedy and projective model category structures on simplicial spaces. However, we need to modify these maps so that they are actually maps between Segal precategories. The purpose of this section is to define two methods of modifying the generating cofibrations and generating acyclic cofibrations so that they are actually maps between Segal prec-ategories, and to prove a result which we need to prove the existence of the model structures $SeCa_{tc}$ and $SeCa_{tf}$.

The first method we call reduction, and we use it to define the gen-erating cofibrations in $SeCa_{tc}$. Consider the forgetful functor from the category of Segal precategories to the category of simplicial spaces.

This map has a left adjoint, which we call the reduction map. Given a simplicial space X, we denote its reduction by $(X)_r$. The degree n space of $(X)_r$ is obtained from X_n by collapsing the subspace $s_0^n x_0$ of X_n to the discrete space $\pi_0(s_0^n x_0)$, where s_0^n is the iterated degeneracy map.

Recall that the cofibrations in the Reedy model category structure on simplicial spaces are monomorphisms (Proposition 2.14) and that the Reedy generating cofibrations are of the form

$$\dot{\Delta}[m] \times \Delta[n]^t \cup \Delta[m] \times \dot{\Delta}[n]^t \to \Delta[m] \times \Delta[n]^t$$

for all n,m≥0. In general, these maps are not in SeCat because the objects involved are not Segal precategories. Therefore, we apply this reduction functor to these maps.

Thus, we consider the maps

$$(\dot{\Delta}[m] \times \Delta[n]^t \cup \Delta[m] \times \dot{\Delta}[n]^t)_r \to (\Delta[m] \times \Delta[n]^t)_r.$$

However, we still need to make some modifications to assure that all these maps are actually monomorphisms. In particular, we need to check the case where n=0. If n=m=0, and if φ denotes the empty simplicial space, we obtain the map φ→Δ[0]t, which is a monomorphism. However, when n=0 and m=1, we get the map $\Delta[0]^t \coprod \Delta[0]^t \to \Delta[0]^t$, which is not a monomorphism. When n=0 and m≥2, we obtain the map $\Delta[0]^t \to \Delta[0]^t$. This map is an isomorphism, and thus there is no reason to include it in the generating set. Therefore, we define the set

$$I_c = \{(\dot{\Delta}[m] \times \Delta[n]^t \cup \Delta[m] \times \dot{\Delta}[n]^t)_r \to (\Delta[m] \times \Delta[n]^t)_r\}$$

for all m≥0 when n≥1 and for n=m=0. This set Ic will be a set of generating cofibrations of SeCat$_c$.

This reduction process works well in almost all situations, but we have problems when we try to reduce some of the generating cofibrations in $\text{SSets}_f^{\Delta^{op}}$, namely the maps

$$\dot{\Delta}[1] \times \Delta[n]^t \to \Delta[1] \times \Delta[n]^t$$

for any $n \geq 0$. The object $\Delta[1] \times \Delta[n]t$ reduces to a Segal precategory with $n+1$ points in degree zero, but the object $\dot{\Delta}[1] \times \Delta[n]^t$ reduces to a Segal precategory with $2(n+1)$ points in degree zero. In other words, the reduced map in this case is no longer a monomorphism.

Consider the set $\Delta[n]_0$ and denote by $\Delta[n]_0^t$ the doubly constant simplicial space defined by it. For $m \geq 1$ and $n \geq 0$, define $P_{m,n}$ to be the pushout of the diagram

$$
\begin{array}{ccc}
\dot{\Delta}[m] \times \Delta[n]_0^t & \longrightarrow & \dot{\Delta}[m] \times \Delta[n]^t \\
\downarrow & & \downarrow \\
\Delta[n]_0^t & \longrightarrow & P_{m,n}.
\end{array}
$$

If $m=0$, then we define $P_{m,0}$ to be the empty simplicial space. For all $m \geq 0$ and $n \geq 1$, define $Q_{m,n}$ to be the pushout of the diagram

$$
\begin{array}{ccc}
\Delta[m] \times \Delta[n]_0^t & \longrightarrow & \Delta[m] \times \Delta[n]^t \\
\downarrow & & \downarrow \\
\Delta[n]_0^t & \longrightarrow & Q_{m,n}.
\end{array}
$$

For each m and n, the map $\dot{\Delta}[m] \times \Delta[n]^t$ induces a map $i_{m,n}: P_{m,n} \to Q_{m,n}$. We then define the set $If = \{i_{m,n}: P_{m,n} \to Q_{m,n} \mid m,n \geq 0\}$. Note that when $m \geq 2$

this construction gives exactly the same objects as those given by re-duction, namely that $P_{m,n}$ is precisely $\dot{\Delta}[m] \times \Delta[n]^t)_r$ and likewise $Q_{m,n}$ is precisely $(\Delta[m] \times \Delta[n]^t)_r$.

Given a Segal precategory X, we denote by $X_n(v_0, \ldots v_n)$ the fiber of the map $X_n \to X_0^{n+1}$ over $(v_0 \ldots v_n) \in X_0^{n+1}$, where this map is given by iter-ated face maps of X. More specifically, $X_0^{n+1} = (\cos k_0 X)_n$ and the map $X_n \to X_0^{n+1}$ is given by the map $X \to \cos k_0 X$.

If Hom denotes morphism set and X is an arbitrary simplicial space, notice that we can use the pushout diagrams defining the objects $P_{m,n}$ and $Q_{m,n}$ to see that

$$\text{Hom}(P_{m,n}, X) \cong \coprod_{v_0, \ldots, v_n} \text{Hom}(\dot{\Delta}[m], X_n(v_0, \ldots, v_n))$$

and

$$\text{Hom}(Q_{m,n}, X) \cong \coprod_{v_0, \ldots, v_n} \text{Hom}(\Delta[m], X_n(v_0, \ldots, v_n)).$$

We now state and prove a lemma using the maps in If.

Lemma 4.1: Suppose a map f: X→Y has the right lifting property with respect to the maps in If. Then the map $X_0 \to Y_0$ is surjective and each map

$$X_n(v_0, \ldots, v_n) \to Y_n(f v_0, \ldots, f v_n)$$

is an acyclic fibration of simplicial sets for each n≥1 and $(v_0 \ldots v_n) \in X_0^{n+1}$.

Proof: The surjectivity of $X_0 \to Y_0$ follows from the fact that f has the right lifting property with respect to the map $P_{0,0} \to Q_{0,0}$.

In order to prove the remaining statement, it suffices to show that there is a dotted arrow lift in any diagram of the form

$$
\begin{array}{ccc}
\dot\Delta[m] & \longrightarrow & X_n(v_0, \ldots, v_n) \\
\downarrow & \nearrow & \downarrow \\
\Delta[m] & \longrightarrow & Y_n(f v_0, \ldots, f v_n)
\end{array}
\tag{4.2}
$$

for $m, n \geq 0$.

By our hypothesis, there is a dotted arrow lift in diagrams of the form

$$
\begin{array}{ccc}
P_{m,n} & \longrightarrow & X \\
\downarrow & \nearrow & \downarrow \\
Q_{m,n} & \longrightarrow & Y
\end{array}
\tag{4.3}
$$

for all $m, n \geq 0$. The existence of the lift in diagram (4.3) is equivalent to the surjectivity of the map $\mathrm{Hom}(Q_{m,n}, x) \to P$ in the following diagram, where P denotes the pullback and Hom denotes morphism set:

$$
\begin{array}{ccccc}
\mathrm{Hom}(Q_{m,n}, X) & \longrightarrow & P & \longrightarrow & \mathrm{Hom}(P_{m,n}, X) \\
& & \downarrow & & \downarrow \\
& & \mathrm{Hom}(Q_{m,n}, Y) & \longrightarrow & \mathrm{Hom}(P_{m,n}, Y).
\end{array}
$$

Now, as noted above we have that

$$
\mathrm{Hom}(Q_{m,n}, X) \cong \coprod_{v_0, \ldots, v_n} \mathrm{Hom}(\Delta[m], X_n(v_0, \ldots, v_n))
$$

and analogous weak equivalences for the other objects of the diagram.

Using these weak equivalences and being particularly careful in the cases where m=1 and m=0, one can show that for each m,n≥0 the dotted arrow lift in diagram (4.2) exists and therefore that each map

$$X_n(v_0, \ldots, v_n) \to Y_n(f v_0, \ldots, f v_n)$$

is an acyclic fibration of simplicial sets for each n≥1.

A SEGAL CATEGORY MODEL CATEGORY STRUCTURE ON SEGAL PRECATEGORIES

In this section, we prove the existence of the model category structure $SeCa_{tc}$.

We would like to define a functorial "localization" functor L_c on Se-Cat such that, given any Segal precategory X, its localization $L_c X$ is a Segal space which is a Segal category weakly equivalent to X in $SeSp_c$. We begin by considering a functorial localization functor in $SeSp_c$ and then modifying it so that it takes values in SeCat. In the case of $SeSp_{c'}$ this localization functor is precisely the functorial fibrant replacement functor.

A choice of generating acyclic cofibrations for $SeSp_c$ is the set of maps

$$V[m, k] \times \Delta[n]^t \cup \Delta[m] \times G(n)^t \to \Delta[m] \times \Delta[n]^t \text{ for } n \geq 0, \ m \geq 1,$$

and 0≤k≤m [12, Section 4.2]. Therefore, one can use the small object argument to construct a functorial localization functor by taking a co-limit of pushouts, each of which is along the coproduct of all these maps [12, Section 4.3].

If we apply this functor to a Segal precategory, the maps with n=0 are problematic because taking pushouts along them will not result in a space which is discrete in degree zero. We claim that we can obtain a functorial localization functor L_c on the category SeCat by taking a colimit of iterated pushouts along the maps

$$V[m, k] \times \Delta[n]^t \cup \Delta[m] \times G(n)^t \rightarrow \Delta[m] \times \Delta[n]^t$$

for n,m≥1 and 0≤k≤m.

To see that this restricted set of maps is sufficient, consider a Segal precategory X and the Segal category L_cX we obtain from taking such a colimit. Then for any 0≤k≤m, consider the diagram

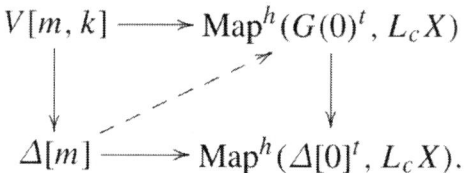

Since $\Delta[0]^t$ is isomorphic to $G(0)^t$, and since L_cX is discrete in degree zero, the right-hand vertical map is an isomorphism of discrete simplicial sets. Therefore, a dotted arrow lift exists in this diagram. It follows that the map $L_cX \rightarrow \Delta[0]^t$ has the right lifting property with respect to the maps

$$V[m, k] \times \Delta[n]^t \cup \Delta[m] \times G(n)^t \rightarrow \Delta[m] \times \Delta[n]^t$$

for all n≥0, m≥1, and 0≤k≤m. Therefore, L_cX is fibrant in $SeSp_c$, namely, a Segal space.

Since L_cX is a Segal space, it makes sense to talk about the mapping space $map_{L_cX}(x,y)$ and the homotopy category $H_o(L_cX)$. Given these facts, we show that there exists a model category structure SeCatc on Segal precategories with the following three distinguished classes of morphisms:

— Weak equivalences are the maps f: X→Y such that the induced map $L_cX \rightarrow L_cY$ is a DK-equivalence of Segal spaces. (Again, we will call such maps DK-equivalences.)
— Cofibrations are the monomorphisms. (In particular, every Segal precategory is cofibrant.)

— Fibrations are the maps with the right lifting property with respect to the maps which are both cofibrations and weak equivalences.

Theorem 5.1: There is a cofibrantly generated model category structure $SeCat_c$ on the category of Segal precategories with the above weak equivalences, fibrations, and cofibrations.

We first need to define sets I_c and J_c as our candidates for generating cofibrations and generating acyclic cofibrations, respectively.

We take as generating cofibrations the set

$$I_c = \{(\dot{\Delta}[m] \times \Delta[n]^t \cup \Delta[m] \times \dot{\Delta}[n]^t)_r \rightarrow (\Delta[m] \times \Delta[n]^t)_r\}$$

for all $m \geq 0$ when $n \geq 1$ and for $n=m=0$. Notice that since taking a push-out along such a map amounts to attaching an m-simplex to the space in degree n, any cofibration can be written as a directed colimit of pushouts along the maps of I_c.

We then define the set $J_c = \{i:A \rightarrow B\}$ to be a set of representatives of isomorphism classes of maps in SeCat satisfying two conditions:

— For all $n \geq 0$, the spaces A_n and B_n have countably many simplices.
— The map $i: A \rightarrow B$ is a monomorphism and a weak equivalence.

Given these proposed generating acyclic cofibrations, we need to show that any acyclic cofibration in $SeCat_c$ is a directed colimit of pushouts along such maps. To prove this result, we require several lemmas. The proofs of the first three we omit here; proofs can be found in the author's thesis [6].

Lemma 5.2: Let A→B be a CW-inclusion. The following statements are equivalent:

— A→B is a weak equivalence of topological spaces.
— For all $n \geq 1$, any map of pairs $(D^n, S^{n-1}) \rightarrow (B, A)$ extends over the map of cones (CD^n, CS^{n-1}).

— For all $n \geq 1$, any map $(D^n, S^{n-1}) \to (B, A)$ is homotopic to a constant map.

Lemma 5.3: Let $f: X \to Y$ be a an inclusion of simplicial sets which is a weak equivalence, and let W and Z be simplicial sets such that we have a diagram of inclusions

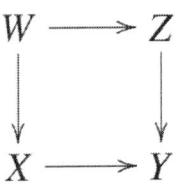

Let $u: (D^n, S^{n-1}) \to (|Z|, |W|)$ be a relative map of CW-pairs. Then the inclusion $i: (|Z|, |W|) \to (|Y|, |X|)$ can be factored as a composite

$$(|Z|, |W|) \to (|K|, |L|) \to (|Y|, |X|)$$

where K is a subspace of Y obtained from Z by attaching a finite number of nondegenerate simplices, L is a subspace of X, and the composite map of relative CW-complexes

$$(D^n, S^{n-1}) \to (|Z|, |W|) \to (|K|, |L|)$$

is homotopic rel S^{n-1} to a map $D^n \to |L|$.

Lemma 5.4: Let (Y, X) be a CW-pair such that X and Y have only countably many cells. Then for a fixed $n \geq 0$, there are only countably many homotopy classes of maps $(D^n, S^{n-1}) \to (Y, X)$.

If $A \to B$ is a monomorphism of Segal precategories, then taking the localization via the small object argument gives us that $L_c A \to L_c B$ is a monomorphism of Segal categories. In particular, if $A \subseteq B$ is an inclusion, then we can regard $L_c A \subseteq L_c B$ as an inclusion as well.

Lemma 5.5: Let A and B be Segal precategories such that $A \subseteq B$. Let σ be a simplex in $L_c B$ which is not in $L_c A$. Then there exists a Segal precategory A′ such that A′ is obtained from A by attaching a finite number of nondegenerate simplices and σ is in LA′.

Proof: By our description of our localization functor at the beginning of the section, L_cB is obtained from B by taking a colimit of pushouts, each of which is along the map

$$\coprod_{m,k,n} V[m, k] \times \Delta[n]^t \cup \Delta[m] \times G(n)^t \rightarrow \coprod_{m,k,n} \Delta[m] \times \Delta[n]^t$$

for $n,m \geq 1$ and $0 \leq k \leq m$. The Segal category L_cB is the colimit of a filtration

$$B \subseteq \Psi^1 B \subseteq \Psi^2 B \subseteq \cdots$$

where each Ψ_i is given by a colimit of iterated pushouts along this map. Since σ is a single simplex, it is small and therefore σ is in $\Psi^n B$ for some n.

Therefore, σ is obtained by attaching $\Delta[m] \times \Delta[n]^t$ along a finite number of non-degenerate simplices of $\Psi^{n-1}B$. We can then apply the preceding argument to each of these simplices and inductively obtain a finite number of non-degenerate simplices of B which form a sub-Segal precategory which we will call C. We then define A′=A∪C.

We then state one more lemma, which is a generalization of a lemma given by Hirschhorn [12, 2.3.6].

Lemma 5.6: Let the map g: A→B be an inclusion of Segal precategories, each of which has countably many simplices. If X is a Segal precategory with countably many simplices, then its localization L_X with respect to the map ghas only countably many simplices.

We are now able to state and prove our result about generating cofibrations.

Proposition 5.7: Any acyclic cofibration $j: C \to D$ in $SeCat_c$ can be written as a directed colimit of pushouts along the maps in J_c

Proof: Note that by definition $j: C \to D$ is a monomorphism of Segal precategories. We assume that it is an inclusion. Let U be a subsimplicial space of D such that U has countably many simplices in each degree. Apply the localization functor L_c to obtain a diagram of Segal categories

$$
\begin{array}{ccc}
L_c(U \cap C) & \longrightarrow & L_c U \\
\downarrow & & \downarrow \\
L_c C & \longrightarrow & L_c D.
\end{array}
$$

Since U has only countably many simplices, this localization process adds at most a countable number of simplices to the original simplicial space by Lemma 5.6.

We would like to find a Segal precategory W such that $U \subseteq W \subseteq D$ and such that the map $W \cap C \to W$ is in the set J_c.

First consider the map

$$
Ho(L_c(U \cap C)) \to Ho(L_c U)
$$

which we want to be an equivalence of categories. If it is not an equivalence, then there exists $z \in (L_c U)_0$ which is not equivalent to some $z' \in (L_c(U \cap C))_0$. However, there is such a z' when we consider z as an element of $(L_c D)_0$, since $j: C \to D$ is a DK-equivalence. If this z' is not in $(U \cap C)_0$, then we add it. Repeat this process for all such z.

Now for each such z, consider the four mapping spaces in $L_c U$ involving the objects z and z':, and $map_{L_cU}(z,z)$, $map_{L_cU}(z,z')$ and $map_{L_cU}(z',z)$,, We want the sets of components of these four spaces to be isomorphic to one another in $H_0(L_c U)$. We can attach a countable number of simplices via an analogous argument to the one in the proof of Lemma 5.5 such that these sets of components are isomorphic. We then repeat the same argument to assure that $\pi_0 \mathrm{map}_{L_cU}(x, y)$ is isomorphic to $\pi_0 \mathrm{map}_{L_cU}(x, z')$ for each $x \in U_0$ and analogously for the sets of components of the mapping spaces out of each such x.

By repeating this process for each such z, we obtain a Segal precategory Y with a countable number of simplices such that $H_0(L_c(Y \cap C)) \to H_0(L_cY)$ is an equivalence of categories. However, we do not necessarily have that for each $x, y \in L_c(Y \cap C)$,

$$\mathrm{map}_{L_c(Y \cap C)}(x, y) \to \mathrm{map}_{L_cY}(x, y)$$

is a weak equivalence of simplicial sets. Therefore we consider all maps

$$(D^n, S^{n-1}) \to (|\mathrm{map}_{L_cY}(x, y)|, |\mathrm{map}_{L_c(Y \cap C)}(x, y)|) \to (|\mathrm{map}_{L_cD}(x, y)|, |\mathrm{map}_{L_cC}(x, y)|)$$

for each $x, y \in (Y \cap C)_0$ and $n \geq 0$. Identify all x, y, and n such that the map

$$(D^n, S^{n-1}) \to (|\mathrm{map}_{L_cY}(x, y)|, |\mathrm{map}_{L_c(Y \cap C)}(x, y)|)$$

is not homotopic to a constant map.

However each composite map

$$(D^n, S^{n-1}) \to (|\mathrm{map}_{L_cY}(x, y)|, |\mathrm{map}_{L_c(Y \cap C)}(x, y)|) \to (|\mathrm{map}_{L_cD}(x, y)|, |\mathrm{map}_{L_cC}(x, y)|)$$

is homotopic to a constant map by Lemma 5.2 since

$$|\mathrm{map}_{L_cC}(x, y)| \to |\mathrm{map}_{L_cD}(x, y)|$$

is a weak equivalence.

For each such x, y, and n, it follows from Lemma 5.3 that there exists some pair of simplicial sets

$$(\mathrm{map}_{L_cY}(x, y), \mathrm{map}_{L_c(Y \cap C)}(x, y)) \subseteq (K, L) \subseteq (\mathrm{map}_{L_cD}(x, y), \mathrm{map}_{L_cC}(x, y))$$

such that the composite map

$$(D^n, S^{n-1}) \to (|\mathrm{map}_{L_cY}(x, y)|, |\mathrm{map}_{L_c(Y \cap C)}(x, y)|) \to (|K|, |L|)$$

is homotopic to a constant map, and the pair (K,L) is obtained from the pair $(\mathrm{map}_{L_cY}(x, y), \mathrm{map}_{L_c(Y \cap C)}(x, y))$ by attaching a finite number of non-degenerate simplices. We apply Lemma 5.5 to each of these new simplices obtained by considering each nontrivial homotopy class to obtain some Segal precategory Y' with a countable number of number of simplices such that each composite map

$$(D^n, S^{n-1}) \to (|\mathrm{map}_{L_cY}(x, y)|, |\mathrm{map}_{L_c(Y \cap C)}(x, y)|) \to (|\mathrm{map}_{L_cY'}(x, y)|, |\mathrm{map}_{L_c(Y' \cap C)}(x, y)|)$$

is homotopic to a constant map.

However, the process of adding simplices may have created more maps

$$(D^n, S^{n-1}) \to (|\mathrm{map}_{L_cY'}(x, y)|, |\mathrm{map}_{L_c(Y' \cap C)}(x, y)|)$$

that are not homotopic to a constant map. Therefore we repeat this argument, perhaps countably many times, until, taking a colimit over all of them, we obtain a Segal precategory W such that each map

$$(D^n, S^{n-1}) \to (|\mathrm{map}_{L_cW}(x, y)|, |\mathrm{map}_{L_c(W \cap C)}(x, y)|)$$

is homotopic to a constant map. Since each of these steps added only countably many simplices to the original Segal precategory U, and since by Lemma 5.2

$$\mathrm{map}_{L_c(W \cap C)}(x, y) \to \mathrm{map}_{L_cW}(x, y)$$

is a weak equivalence for all $x, y \in (L_c(W \cap C))_0$, the map $W \cap C \to W$ is in the set J_c.

Now, take some $\tilde{U}$ obtained from W by adding a countable number of simplices, consider the inclusion map $\tilde{U} \cap C \to \tilde{U}$, and repeat the entire process. To show that we can repeat this argument, taking a (possibly transfinite) colimit, and eventually obtain the map j: C→D, it suffices to show that the localization functor L_c commutes with arbitrary directed colimits of inclusions. However, this fact follows from [12, 2.2.18].

Now, we have two definitions of acyclic fibration that we need to show coincide: the fibrations which are weak equivalences, and the maps with the right lifting property with respect to the maps in I_c.

Proposition 5.8: The maps with the right lifting property with respect to the maps in Icare fibrations and weak equivalences.

Before giving a proof of this proposition, we begin by looking at the maps in I_c and determining what an I_c-injective looks like. Recall the definition of the coskeleton of a simplicial space from the paragraph following Proposition 2.14. If f: X→Y has the right lifting property with respect to the maps in $I_{c'}$, then for each n≥1, the map $X_n \to P_n$ is an acyclic fibration of simplicial sets, where P_n is the pullback in the diagram

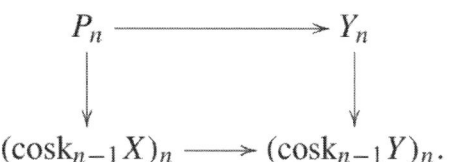

$$P_n \longrightarrow Y_n$$
$$\downarrow \qquad\qquad \downarrow$$
$$(\mathrm{cosk}_{n-1}X)_n \longrightarrow (\mathrm{cosk}_{n-1}Y)_n.$$

In the case that n=0, the restrictions on m and n give us that the map $X_0 \to Y_0$ is a surjection rather than the isomorphism we get in the Reedy case. Notice that by the same argument given for the Reedy model category structure (in the section following Proposition 2.14 above), the simplicial sets P_n can be characterized up to weak equivalence as homotopy pullbacks and are therefore homotopy invariant.

This characterization of the maps with the right lifting property with respect to I_c will enable us to prove Proposition 5.8. Before proceeding to the proof, however, we state a lemma, whose proof we defer to Section 9.

Lemma 5.9: Suppose that f: X→Y is a map of Segal precategories which is an I_c-injective. Then f is a DK-equivalence.

Proof of Proposition 5.8: Suppose that f: X→Y is an I_c-injective, or a map which has the right lifting property with respect to the maps in I_c. Note that f then has the right lifting property with respect to all cofibrations. Since, in particular, it has the right lifting property with respect to the acyclic cofibrations, it is a fibration by definition. It remains to show that f is a weak equivalence.

However, this fact follows from Lemma 5.9, proving the proposition.

We now state the converse, which we prove in Section 9.

Proposition 5.10: The maps in $SeCat_c$ which are both fibrations and weak equivalences are I_c-injectives.

Now we prove a lemma which we need to check the last condition for our model category structure.

Lemma 5.11: A pushout along a map of J_c is also an acyclic cofibration in $SeCat_c$.

Proof: Let j: A→B be a map in J_c. Notice that j is an acyclic cofibration in the model category CSS. Since CSS is a model category, we know that a pushout along an acyclic cofibration is again an acyclic cofibration [10, 3.14(ii)]. If all the objects involved are Segal precategories, then the pushout will again be a Segal precategory and therefore the pushout map will be an acyclic cofibration in $SeCat_c$.

Proposition 5.12: If a map of Segal precategories is a J_c-cofibration, then it is an I_c-cofibration and a weak equivalence.

Proof: By definition and Proposition 5.7, a J_c-cofibration is a map with the left lifting property with respect to the maps with the right lifting property with respect to the acyclic cofibrations. However, by the definition of fibration, these maps are the ones with the left lifting property with respect to the fibrations.

Similarly, using Proposition 5.8 and Proposition 5.10, an I_c-cofibration is a map with the left lifting property with respect to the acyclic fibrations. Thus, we need to show that a map with the left lifting property with respect to the fibrations has the left lifting property with respect to the acyclic fibrations and is a weak equivalence. Since the acyclic fibrations are fibrations, it remains to show that the maps with the left lifting property with respect to the fibrations are weak equivalences.

Let f: A→B be a map with the left lifting property with respect to all fibrations. By Lemma 5.11 above, we know that a pushout along maps of Jc is an acyclic cofibration. Therefore, we can use the small object argument [12, 10.5.15] to factor the map f: A→B as the composite of an acyclic cofibration A→A' and a fibration A'→B. Then there exists a dotted arrow lift in the diagram

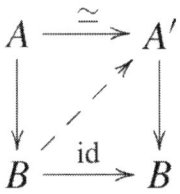

showing that the map A→B is a retract of the map A→A' and therefore a weak equivalence.

Proof of Theorem 5.1: Axiom MC_1 follows since limits and colimits of Segal precategories (computed as simplicial spaces) still have discrete zero space and are therefore Segal precategories. MC_2 and MC_3 (for weak equivalences) work as usual, for example see [10, 8.10].

It remains to show that the four conditions of Theorem 2.3 are satisfied. The set I_c permits the small object argument because the generating cofibrations in the Reedy model category structure do. We can show that the objects A which appear as the sources of the maps in J_c are small using an analogous argument to the one for simplicial sets [12,

10.4.4], so the set J_c permits the small object argument. Thus, condition 1 is satisfied.

Condition 2 is precisely the statement of Proposition 5.12. Condition 3 and condition 4(ii) are precisely the statements of Proposition 5.8 and Proposition 5.10.

Note that the reduced Reedy acyclic cofibrations

$$(V[m,k] \times \Delta[n]^t \cup \Delta[m] \times \dot{\Delta}[n]^t)_r \rightarrow (\Delta[m] \times \Delta[n]^t)_r$$

are acyclic cofibrations in SeCatc for m≥0 when n≥1 and for n=m=0.

Corollary 5.13: The fibrant objects in $SeCat_c$ are Reedy fibrant Segal categories.

Proof: Suppose that X is fibrant in $SeCat_c$. Then, since the reduced Reedy cofibrations are acyclic cofibrations in $SeCat_c$ and since X has discrete zero space, it follows that X is Reedy fibrant.

Then, since the maps

$$(\Delta[m] \times G(n)^t)_r \rightarrow (\Delta[m] \times \Delta[n]^t)_r$$

for all m,n≥0 are acyclic cofibrations in $SeCat_c$, it follows that X is a Segal category.

The converse statement, that the Reedy fibrant Segal categories are fibrant in $SeCat_c$, also holds [2].

A QUILLEN EQUIVALENCE BETWEEN SECATC AND CSS

In this section, we will show that there is a Quillen equivalence between the model category structure $SeCat_c$ on Segal precategories and the complete Segal space model category structure CSS on simplicial spaces. We first need to show that we have an adjoint pair of maps between the two categories.

Let I: $SeCat_c \to CSS$ be the inclusion functor of Segal precategories into the category of all simplicial spaces. We will show that there is a right adjoint functor R: $CSS \to SeCat_c$ which "discretizes" the degree zero space.

Let W be a simplicial space. Define simplicial spaces $U=cosk_0(w_0)$ and. $V=cosk_0(w_{0,0})$There exist maps $W \to U \leftarrow V$. Then we take the pullback RW in the diagram

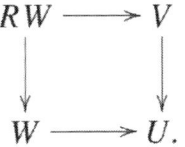

Note that RW is a Segal precategory. If W is a complete Segal space, then so are U and V, and in this case RW is a Segal category, which we can see as follows. The pullback at degree 1 gives

$$
\begin{array}{ccc}
(RW)_1 & \longrightarrow & W_{0,0} \times W_{0,0} \\
\downarrow & & \downarrow \\
W_1 & \longrightarrow & W_0 \times W_0
\end{array}
$$

and at degree 2 we get

$$
\begin{array}{ccc}
(RW)_1 \times_{(RW)_0} (RW)_1 & \longrightarrow & (W_{0,0})^3 \\
\downarrow & & \downarrow \\
W_2 \simeq W_1 \times_{W_0} W_1 & \longrightarrow & W_0 \times W_0 \times W_0.
\end{array}
$$

Looking at these pullbacks, and the analogous ones for higher n, we notice that RW is in fact a Segal category.

We define the functor R: CSS→SeCat$_c$ which takes a simplicial space W to the Segal precategory RW given by the description above.

Proposition 6.1: The functor R: CSS→SeCat$_c$ is right adjoint to the inclusion map I: SeCat$_c$→CSS.

Proof: We need to show that there is an isomorphism

$$\text{Hom}_{\mathcal{S}e\mathcal{C}at_c}(Y, RW) \cong \text{Hom}_{\mathcal{C}SS}(IY, W)$$

for any Segal precategory Y and simplicial space W.

Suppose that we have a map Y=IY→W. Since Y is a Segal precategory, Y_0 is equal to $Y_{0,0}$ viewed as a constant simplicial set. Therefore, we can restrict this map to a unique map Y→V, where V is the Segal precategory defined above. Then, given the universal property of pullbacks, there is a unique map Y→RW. Hence, we obtain a map

$$\varphi: \text{Hom}_{\mathcal{C}SS}(IY, W) \to \text{Hom}_{\mathcal{S}e\mathcal{C}at_c}(Y, RW).$$

This map is surjective because given any map Y→RW we can compose it with the map RW→W to obtain a map Y→W.

Now for any Segal precategory Y, consider the diagram

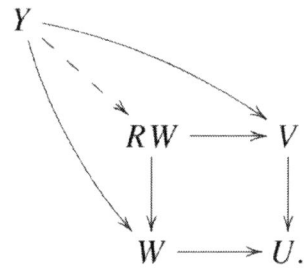

Because this diagram must commute and the image of the map $Y_0 \to W_0$ is contained in $W_{0,0}$ since Y is a Segal precategory, this map uniquely determines what the map $Y \to V$ has to be. Therefore, given a map $Y \to RW$, it could only have come from one map $Y \to W$. Thus, φ is injective.

Now, we need to show that this adjoint pair respects the model category structures that we have.

Proposition 6.2: The adjoint pair of functors

$$I : Se\mathcal{C}at_c \rightleftarrows \mathcal{C}SS : R$$

— is a Quillen pair.

Proof: It suffices to show that the inclusion map I preserves cofibrations and acyclic cofibrations. I preserves cofibrations because they are defined to be monomorphisms in each category. Also in each of the two categories, a map is a weak equivalence if it is a DK-equivalence after localizing to obtain a Segal space, as given in Theorem 3.10. In each case an acyclic cofibration is an inclusion satisfying this property. Therefore, the map I preserves acyclic cofibrations.

Theorem 6.3: The Quillen pair

$$I : Se\mathcal{C}at_c \rightleftarrows \mathcal{C}SS : R$$

is a Quillen equivalence.

Proof: We need to show that I reflects weak equivalences between cofibrant objects and that for any fibrant object W (i.e., complete Segal space) in CSS, the map $I((RW)^c) = IRW \to W$ is a weak equivalence in $SeCat_c$.

The fact that I reflects weak equivalences between cofibrant objects follows from the same argument as the one in the proof of the Quillen pair. To prove the second part, it remains to show that the map j: RW→W in the pullback diagram

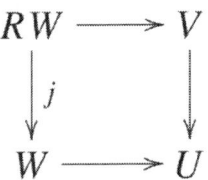

is a DK-equivalence. It suffices to show that the map of objects $op(RW) \to op(W)$ is surjective and that the map $map_{RW}(x, y) \to map_W(jx, jy)$ is a weak equivalence, where the object set of a Segal space is defined as in Section 3.3. However, notice by the definition of RW that ob(RW)=ob(W). In particular, jx=x and jy=y. Then notice, using the pullback that defines $(RW)_1$, that $map_{RW}(x, y) \simeq map_W(x, y)$. Therefore, the map RW→W is a DK-equivalence.

ANOTHER SEGAL CATEGORY MODEL CATEGORY STRUCTURE ON SEGAL PRECATEGORIES

The model category structure $SeCat_c$ that we defined above is helpful for the Quillen equivalence with the complete Segal space model category structure, but there does not appear to be a Quillen equivalence between it and the model category structure SC on simplicial categories. Therefore, we need another model category structure $SeCat_f$ to obtain such a Quillen equivalence.

In the model category structure $SeCat_c$, we started with the generating cofibrations in the Reedy model category structure and adapted them to be generating cofibrations of Segal precategories. In this second model category structure, we use modified generating cofibrations from the projective model category structure on simplicial spaces so that the objects involved are Segal precategories.

We make the following definitions for a model category structure Se-Cat$_f$ on the category of Segal precategories.

- The weak equivalences are the same as those of SeCat$_c$.
- The cofibrations are the maps which can be formed by taking iterated pushouts along the maps of the set if defined in Section 4.
- The fibrations are the maps with the right lifting property with respect to the maps which are cofibrations and weak equivalences.

Notice that to define the weak equivalences in this case we want to use a functorial localization in SeSp$_f$ rather than SeSp$_c$. We define a localization functor Lf in the same way that we defined L$_c$ at the beginning of Section 5 but making necessary changes in light of the fact that we are starting from the model structure SeSp$_f$. So, in a sense, the weak equivalences are not defined identically in the two categories, since they make use of the same localization of different model category structures on the category of simplicial spaces. However, in each case the weak equivalences are the same in the unlocalized model category, so we can define homotopy function complexes using only the underlying category and the weak equivalences. Recall by the definition of local objects that a map X→Y is a local equivalence if and only if the induced map of homotopy function complexes

$$\mathrm{Map}^h(Y, Z) \to \mathrm{Map}^h(X, Z)$$

is a weak equivalence of simplicial sets for any local Z. In particular, the weak equivalences of the localized category depend only on the weak equivalences of the unlocalized category. Therefore the weak equivalences in SeCat$_c$ and SeCat$_f$ are actually the same.

Theorem 7.1: There is a cofibrantly generated model category structure SeCat$_f$ on the category of Segal precategories in which the weak equivalences, fibrations, and cofibrations are defined as above.

We define the set J$_f$ to be a set of isomorphism classes of maps {i:A→B} such that

— for all $n \geq 0$, the spaces An and Bn have countably many simplices, and
— i: A→B is an acyclic cofibration.

We would like to show that If (defined in Section 4) is a set of generating cofibrations and that J_f is a set of generating acyclic cofibrations for SeCat$_f$.

We begin with the following lemma.

Lemma 7.2: Any acyclic cofibration j: C→D in SeCatf$_c$ an be written as a directed colimit of pushouts along the maps in J_f.

Proof: The argument that we used to prove Proposition 5.7 still holds, applying the functor L_f rather than L_c.

Proposition 7.3: A map f: X→Y is an acyclic fibration in SeCatfif and only if it is an If-injective.

Proof: First suppose that f has the right lifting property with respect to the maps in If. Then we claim that for each $n \geq 0$ and $(v_0 \ldots \ldots v_n) \in X_0^{n+1}$, the map $X_n(v_0, \ldots, v_n) \to Y_n(f_{v_0}, \ldots, f_{v_n})$ is an acyclic fibration of simplicial sets. This fact, however, follows from Lemma 4.1. In particular, it is a weak equivalence, and therefore we can apply the proof of Lemma 5.9 to show that the map X→Y is a DK-equivalence, completing the proof of the first direction. (The proof does not follow precisely in this case, in particular because not all monomorphisms are cofibrations. However, we can use the fact that weak equivalences are the same in SeCat$_c$ and SeCat$_f$ to see that the argument still holds.)

Then, to prove the converse, assume that f is a fibration and a weak equivalence. Then we can apply the proof of Proposition 5.10, making the factorizations in the projective model category structure rather than in the Reedy model category structure. The argument follows analogously.

Proposition 7.4: A map in SeCat$_f$ is a J_f-cofibration if and only if it is an If-cofibration and a weak equivalence.

Proof: This proof follows just as the proof of Proposition 5.12, again using the projective structure rather than the Reedy structure.

Proof of Theorem 7.1: As before, we must check the conditions of Theorem 2.3. Condition 1 follows just as in the proof of Theorem 5.1. Condition 2 is precisely the statement of Proposition 7.4. Condition 3 and condition 4(ii) follow from Proposition 7.3 after applying Lemma 7.2.

We now prove that both our model category structures on the category of Segal precategories are Quillen equivalent.

Theorem 7.5: The identity functor induces a Quillen equivalence

$$I : \mathcal{S}e\mathcal{C}at_f \rightleftarrows \mathcal{S}e\mathcal{C}at_c : J.$$

Proof: Since both maps are the identity functor, they form an adjoint pair. We then show that this adjoint pair is a Quillen pair.

We first make some observations between the two categories. Notice that the cofibrations of SeCat$_f$ form a subclass of the cofibrations of SeCat$_c$ since they are monomorphisms. Similarly, the acyclic cofibrations of SeCat$_f$ form a subclass of the acyclic cofibrations of SeCat$_c$.

In particular, these observations imply that the left adjoint I:SeCat$_f$→SeCat$_c$ preserves cofibrations and acyclic cofibrations. Hence, we have a Quillen pair.

It remains to show that this Quillen pair is a Quillen equivalence. To do so, we must show that given any cofibrant X in SeCat$_f$ and fibrant Y in SeCatc, a map f: IX→Y is a weak equivalence in SeCat$_f$ if and only if φf: X→JY is a weak equivalence in SeCat$_c$. However, this follows from the fact that the weak equivalences are the same in each category.

Note One might ask at this point why we could not just use the SeCatf model category structure and show a Quillen equivalence between it and the model category structure CSS$_f$ where we localize the projective model category structure (rather than the Reedy) with respect to the maps φ and Ψ. The existence of such a Quillen equivalence would certainly simplify this paper!

However, if we work with "complete Segal spaces" which are fibrant in the projective model structure rather than in the Reedy structure, then for a fibrant object W the map W→U used in defining the right adjoint CSS→SeCat$_c$ is no longer necessarily a fibration. Therefore, the pullback RW is no longer a homotopy pullback and in particular not homotopy invariant. If RW is not homotopy invariant, then there is no guarantee that the map RW→W is a DK-equivalence, and the argument for a Quillen equivalence fails. Thus, the SeCat$_c$ and CSS model structures are necessary.

A QUILLEN EQUIVALENCE BETWEEN Sc AND SeCat$_f$

We begin, as above, by defining an adjoint pair of functors between the two categories SC and SeCat$_f$. We have the nerve functor R: SC→SeCatf. In order to define a left adjoint to this map, we need some terminology.

Definition 8.1: Let D be a small category and SSets$_D$ the category of functors D→SSets. Let S be a set of morphisms inSSetsD. An object Y of SSets$_D$ is strictly S-local if for every morphism f: A→B in S, the induced map on function complexes

$$f^* : \mathrm{Map}(B, Y) \to \mathrm{Map}(A, Y)$$

is an isomorphism of simplicial sets. A map g: C→D in SSets$_D$ is a strict S-local equivalence if for every strictly S-local object Y in SSets$_D$, the induced map

$$g^* : \mathrm{Map}(D, Y) \to \mathrm{Map}(C, Y)$$

is an isomorphism of simplicial sets.

Now, we can view Segal precategories as functors $\Delta^{op} \to$ SSets . Because we require the image of [0] to be a discrete simplicial set, the category of Segal precategories is a subcategory of the category of all such functors. In this section, we are going to regard simplicial categories as the strictly local objects in SeCat$_f$ with respect to the map φ described in Section 3.3.

Although we are actually working in a subcategory, we can still use the following lemma to obtain a left adjoint functor F to our inclusion map R, since the construction will always produce a simplicial space with discrete 0-space when applied to such a simplicial space.

Lemma 8.2 [4, 5.6]: Consider two categories, the category of all diagrams X: D→SSetsand the category of strictly local diagrams with respect to the set of maps S={f: A→B}. The forgetful functor from the category of strictly local diagrams to the category of all diagrams has a left adjoint.

We define the functor F: SeCat$_f$→SC to be this left adjoint to the inclusion functor of strictly local diagrams into all diagrams R: SC→SeCat$_f$.

Proposition 8.3: The adjoint pair

$$F : \mathcal{S}e\mathcal{C}at_f \rightleftarrows \mathcal{S}\mathcal{C} : R$$

is a Quillen pair.

Proof: We prove that this adjoint pair is a Quillen pair by showing that the left adjoint F preserves cofibrations and acyclic cofibrations. We begin by considering cofibrations.

Since F is a left adjoint functor, it preserves colimits. Therefore, it suffices to show that F preserves the setIf of generating cofibrations in Se-Cat$_f$. Recall that the elements of this set are the maps Pm,n→Qm,n as defined in Section 4. We begin by considering the maps $P_{n,1} \to Q_{n,1}$ for any n≥0. The strict localization of such a map is precisely the map of simplicial categories $U\dot{\Delta}[n] \to U\Delta[n]$ (Section 3.1) which is a generating cofibration in SC. We can also see that the strict localization of any $P_{m,n} \to Q_{m,n}$ can be obtained as the colimit of iterated pushouts along the generating cofibrations of SC. Therefore, F preserves cofibrations.

We now need to show that F preserves acyclic cofibrations. To do so, first consider the model category structure $LSSets_{O,f}^{\Delta^{op}}$ (defined in Section 3.15) on Segal precategories with a fixed set O in degree zero and the model category structure SC_O of simplicial categories with a fixed object set O. Recall from Section 3.15 that there is a Quillen equivalence

$$F_O : LSSets_{O,f}^{\Delta^{op}} \rightleftarrows SC_O : R_O.$$

In particular, if X is a cofibrant object of $LSSets_{O,f}^{\Delta^{op}}$, then there is a weak equivalence $X \to R_O((F_O X)^f)$. Notice that F_O agrees with F on Segal precategories with the set O in degree zero, and similarly for R_O and R.

Suppose, then, that X is an object of $LSSets_{O,f}^{\Delta^{op}}$, Y is an object of $LSSets_{O,f}^{\Delta^{op}}$, and $X \to Y$ is an acyclic cofibration in $SeCat_f$. We have a commutative diagram

$$
\begin{array}{ccc}
X & \xrightarrow{\sim} & L_f X \\
\downarrow & & \downarrow \\
Y & \xrightarrow{\sim} & L_f Y
\end{array}
$$

where the upper and lower horizontal maps are weak equivalences not only in $SeCat_f$, but in $LSSets_{O,f}^{\Delta^{op}}$ and $LSSets_{O,f}^{\Delta^{op}}$, respectively. However, using the fixed object case Quillen equivalence, the functors F_O and $F_{O'}$ (and hence F) will preserve these weak equivalences, giving us a diagram

$$
\begin{array}{ccc}
FX & \xrightarrow{\sim} & FL_f X \\
\downarrow & & \downarrow \\
FY & \xrightarrow{\sim} & FL_f X.
\end{array}
$$

Using these weak equivalences and our assumption that $L_f X \to L_f Y$ is a DK-equivalence, we obtain a diagram

$$
\begin{array}{ccc}
L_f X & \xrightarrow{\simeq} & RFL_f X \\
{\scriptstyle\simeq}\downarrow & & \downarrow \\
L_f Y & \xrightarrow{\simeq} & RFL_f Y
\end{array}
$$

in which the upper horizontal arrow is a weak equivalence in $\mathrm{LSSets}_{o,f}^{\Delta^{op}}$ and the lower horizontal arrow is a weak equivalence in $\mathrm{LSSets}_{o,f}^{\Delta^{op}}$. The commutativity of this diagram implies that the map $RFL_f X \to RFL_f Y$ is a DK-equivalence also. Thus, we have shown that F preserves acyclic cofibrations between cofibrant objects.

It remains to show that F preserves all acyclic cofibrations. Suppose that f: X→Y is an acyclic cofibration in SeCat_f. Apply the cofibrant replacement functor to the map X→Y to obtain an acyclic cofibration X′→Y′, and notice that in the resulting commutative diagram

$$
\begin{array}{ccc}
X' & \longrightarrow & Y' \\
\downarrow & & \downarrow \\
X & \longrightarrow & Y
\end{array}
$$

the vertical arrows are level wise weak equivalences.

Now consider the following diagram, where the top square is a push out diagram:

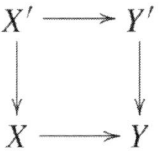

$$
\begin{array}{ccc}
X' & \xrightarrow{\simeq} & Y' \\
\downarrow & & \downarrow \\
X & \xrightarrow{\simeq} & Y'' \\
\shortparallel\downarrow & & \downarrow \\
X & \xrightarrow{\simeq} & Y.
\end{array}
$$

Notice that all three of the horizontal arrows are acyclic cofibrations in SeCat$_f$, the upper and lower by assumption and the middle one because pushouts preserve acyclic cofibrations [10, 3.14]. Now we apply the functor F to this diagram to obtain a diagram

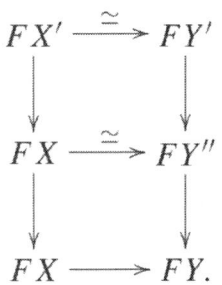

$$(8.4)$$

The top horizontal arrow is an acyclic cofibration since F preserves acyclic cofibrations between cofibrant objects. Furthermore, since F is a left adjoint and hence preserves colimits, the middle horizontal arrow is also an acyclic cofibration because the top square is a push out square.

Now, recall that, given an object X in a model category C, the category of objects under X has as objects the morphisms X→Y in C for any object Y, and as morphisms the maps Y→Y' in C making the appropriate triangular diagram commute [12, 7.6.1]. There is a model category structure on this under category in which a morphism is a weak equivalence, fibration, or cofibration if it is in C [12, 7.6.5]. In particular, a object X→Y is cofibrant in the under category if it is a cofibration in C.

With this definition in mind, to show that the bottom horizontal arrow of diagram (8.4) is an acyclic cofibration, consider the following diagram in the category of cofibrant objects under X:

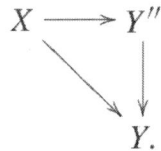

Now, let O'' denote the set in degree zero of Y'' (and also of Y) which is not in the image of the map from X. Now we have the diagram in the category of cofibrant objects under $X \amalg O''$ with the same set in degree zero

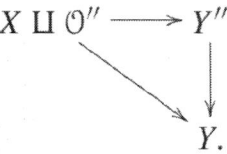

$$X \amalg O'' \longrightarrow Y''$$
$$\searrow \quad \downarrow$$
$$Y.$$

However, since we are now working in a fixed object set situation, we know by Theorem 3.16 that F_O'' is the left adjoint of a Quillen pair, and therefore the map $F_O''Y'' \to F_O''Y$ is a weak equivalence in SC_O'', and in particular a DK-equivalence when regarded as a map in SC. It follows that the map $FX \to FY$ is a weak equivalence, and F preserves acyclic cofibrations.

Recall that we are regarding a Segal category as a local diagram and a simplicial category as a strictly local diagram in $SeCat_f$.

Lemma 8.5: The map $X \to FX$ is a DK-equivalence for every cofibrant object X in $SeCat_f$.

Proof: First consider a free diagram in $SeCat_f$, namely some $\amalg_i Q_{m_i, n_i}$ where each Q_{m_i, n_i} is defined as in Section 4. If Y is a fibrant object in $SeCat_f$, then we have

$$\text{Map}_{\mathcal{S}e\,\mathcal{C}at_f}\left(\coprod_i Q_{m_i,n_i}, Y\right) \simeq \prod_i \text{Map}_{\mathcal{S}e\,\mathcal{C}at_f}(Q_{m_i,n_i}, Y)$$

$$\simeq \prod_i \coprod_{v_0,\dots,v_n} \text{Map}_{\mathcal{S}\mathcal{S}ets}(\Delta[m_i], Y_{n_i}(v_0, \dots, v_n))$$

$$\simeq \prod_i \coprod_{v_0,\dots,v_n} \text{Map}_{\mathcal{S}\mathcal{S}ets}(\Delta[0], Y_{n_i}(v_0, \dots, v_n))$$

$$\simeq \text{Map}_{\mathcal{S}e\,\mathcal{C}at_f}\left(\coprod_i Q_{0,n_i}, Y\right)$$

$$\simeq \text{Map}_{\mathcal{S}e\,\mathcal{C}at_f}\left(\coprod_i \Delta[n_i]^t, Y\right)$$

Therefore, it suffices to consider free diagrams $\coprod_i \Delta[n_i]^t$. Such a diagram is a Segal category. It is also the nerve of a category and thus a strictly local diagram. It follows that the map

$$\coprod_i \Delta[n_i]^t \to F\left(\coprod_i \Delta[n_i]^t\right)$$

is a DK-equivalence.

Now suppose that X is any cofibrant object in SeCat$_f$. Then X can be written as a directed colimit $X \simeq \text{colim}_{\Delta^{op}} X_j$, where each X_j can be written as $\coprod_i \Delta[n_i]^t$. As before we regard FX as a strictly local object in SeCat$_f$. If Y is a fibrant object in SeCat$_f$ which is strictly local, we have

$$\text{Map}_{\mathcal{S}e\,\mathcal{C}at_f}(\text{colim}_{\Delta^{op}} X_j, Y) \simeq \lim_{\Delta} \text{Map}_{\mathcal{S}e\,\mathcal{C}at_f}(X_j, Y)$$

$$\simeq \lim_{\Delta} \text{Map}_{\mathcal{S}e\,\mathcal{C}at_f}(FX_j, Y)$$

$$\simeq \text{Map}_{\mathcal{S}e\,\mathcal{C}at_f}(\text{colim}_{\Delta^{op}} FX_j, Y)$$

$$\simeq \text{Map}_{\mathcal{S}e\,\mathcal{C}at_f}(F(\text{colim}_{\Delta^{op}}(FX_j)), Y).$$

We can now apply the result that

$$F(\mathrm{colim}(FX_j)) \simeq F(\mathrm{colim}X_j).$$

(This fact is proved in [5, 4.1] for ordinary localization, but it holds for strict localization in this case since each X_j is cofibrant and F preserves cofibrant objects.) Therefore we have

$$\mathrm{Map}_{\mathcal{S}e\mathcal{C}at_f}(F(\mathrm{colim}_{\Delta^{op}}(FX_j)), Y) \simeq \mathrm{Map}_{\mathcal{S}e\mathcal{C}at_f}(FX, Y).$$

It follows that the map X→FX is a DK-equivalence.

We are now able to prove the main result of this section.

Theorem 8.6: The Quillen pair

$$F : \mathcal{S}e\mathcal{C}at_f \rightleftarrows \mathcal{S}\mathcal{C} : R$$

is a Quillen equivalence.

Proof: We first show that F reflects weak equivalences between co-fibrant objects. Let f: X→Y be a map of cofibrant Segal precategories such that Ff: FX→FY is a weak equivalence of simplicial categories. (Since F preserves cofibrations, both FX and FY are again cofibrant.) Then consider the following diagram:

$$
\begin{array}{ccccc}
FX & \longrightarrow & L_f FX & \longleftarrow & L_f X \\
{\scriptstyle\simeq}\downarrow & & \downarrow & & \downarrow \\
FY & \longrightarrow & L_f FY & \longleftarrow & L_f Y.
\end{array}
$$

By assumption, the leftmost vertical arrow is a DK-equivalence. The horizontal arrows of the left-hand square are also DK-equivalences by definition. Since X and Y are cofibrant, Lemma 8.5 shows that the horizontal arrows of the right-hand square are DK-equivalences. The commutativity of the whole diagram shows that the map $L_f FX \to L_f FY$ is a DK-equivalence and then that the map $L_f X \to L_f Y$ is also. Therefore, F reflects weak equivalences between cofibrant objects.

Now, we will show that given any fibrant simplicial category Y, the map

$$F((RY)^c) \to Y$$

is a DK-equivalence. Consider a fibrant simplicial category Y and apply the functor R to obtain a Segal category which is level wise fibrant and therefore fibrant in SeCat_f. Its cofibrant replacement will be DK-equivalent to it in SeCat_f. Then, by the above argument, strictly localizing this object will again yield a DK-equivalent simplicial category.

PROOFS OF LEMMA 5.9 AND PROPOSITION 5.10

In this section, we give a proof of two results stated in Section 5. We begin with a lemma which we will use in the proof of Lemma 5.9.

Lemma 9.1: Suppose that f: X→Y is a map of Segal precategories with the right lifting property with respect to the maps in I_c. Then

— The map $f_0 : X_0 \to Y_0$ is surjective, and
— The map $X_n(v_0, \ldots, v_n) \to Y_n(fv_0, \ldots, fv_n)$ is a weak equivalence of simplicial sets for all $n \geq 1$ and $(v_0 \ldots \ldots v_n) \in X_0^{n+1}$.

Proof: Since f: X→Y has the right lifting property with respect to the maps in I_c, it has the right lifting property with respect to all cofibrations. In particular, it has the right lifting property with respect to the maps in the set If. Therefore we can apply Lemma 4.1 and the result follows.

Proof of Lemma 5.9: To prove Lemma 5.9, we consider a given map f: X→Y with the right lifting property with respect to the maps in I_c. It follows from Lemma 9.1 that the map $X_0 \to Y_0$ is surjective and such that for all $n \geq 1$ and $(v_0 \ldots \ldots v_n) \in X_0^{n+1}$ the map

$$X_n(v_0, \ldots, v_n) \to Y_n(f v_0, \ldots, f v_n)$$

is a weak equivalence of simplicial sets.

We must prove that $\mathrm{map}_{L_C X}(x, y) \to \mathrm{map}_{L_C Y}(fx, fy)$ is a weak equivalence of simplicial sets. Given that fact, combining it with the surjectivity of the map $X_0 \to Y_0$ implies that $H_o(L_C X) \to H_o(L_C Y)$ is an equivalence of categories.

We construct a factorization $X \to \Phi Y \to Y$ such that $(\Phi Y)_0 = X_0$ and the map

$$(\Phi Y)_n(v_0, \ldots, v_n) \to Y_n(f v_0, \ldots, f v_n)$$

is an isomorphism of simplicial sets for all $(v_0 \ldots \ldots v_n) \in (\Phi Y)_0^{n+1}$. We begin by defining the object ΦY as the pullback of the diagram

$$\begin{array}{ccc} \Phi Y & \longrightarrow & Y \\ \downarrow & & \downarrow \\ \mathrm{cosk}_0(X_0) & \longrightarrow & \mathrm{cosk}_0(Y_0). \end{array}$$

Note in particular that $(\Phi Y)_0 = X_0$.

Now, notice that for each $n \geq 1$ and $(v_0 \ldots \ldots v_n) \in (\Phi Y)_0^{n+1}$ the map

$$(\Phi Y)_n(v_0, \ldots, v_n) \to Y_n(f v_0, \ldots, f v_n)$$

is an isomorphism of simplicial sets. Since each

$$X_n(v_0, \ldots, v_n) \to Y_n(f v_0, \ldots, f v_n)$$

is a weak equivalence, we can apply model category axiom MC_2 to simplicial sets to see that the map

$$X_n(v_0, \ldots, v_n) \to (\Phi Y)_n(v_0, \ldots, v_n)$$

is a weak equivalence for each n≥1 and $(v_0,\ldots v_n)$ also.

Thus we have shown that if X→Y has the right lifting property with respect to the maps in I_c, then each map $X_n(v_0,\ldots,v_n) \to (\Phi Y)_n (v_0,\ldots,v_n)$ is a weak equivalence of simplicial sets for n≥1 and $(v_0\ldots..v_n) \in X_0^{n+1}$. Since $X_0 = (\Phi Y)_0$, the map X→ΦY is actually a Reedy weak equivalence and therefore also a DK-equivalence. To prove Lemma 5.9, it remains to show that the map ΦY→Y is a DK-equivalence, implying that the map X→Y is also. We will prove this fact by induction on the skeleta of Y.

We will denote by $sk_n Y$ the n-skeleton of Y, as defined above in the paragraph below Proposition 2.14. We seek to prove that the map

$$\Phi(sk_n Y) \to sk_n Y$$

is a DK-equivalence for all n≥0.

We first consider the case where n=0. In this case, $sk_0(\Phi Y)$ and $sk_0 Y$ are already Segal categories. They can be observed to be DK-equivalent as follows. In the case of $sk_0 Y$, given any pair of elements $(x, y) \in (sk_0 Y)_0 \times (sk_0 Y)_0$, the mapping space $map_{sk_0 Y}(x, y)$ is the homotopy fiber of the map

$$(sk_0 Y)_1 = (sk_0 Y)_0 \times (sk_0 Y)_0 \to (sk_0 Y)_0 \times (sk_0 Y)_0$$

over (x,y). If x=y, this fiber is just the point (x,y), since in this case this map is the identity. If x≠y, then the fiber is empty. For $(a,b) \in (sk_0 \Phi Y)_0 \times (sk_0 \Phi Y)_0$, the fiber of the analogous map over (a,b) is equivalent to (a,b) if a and b map to the same point x in Y_0. Otherwise the fiber is empty. The definition of ΦY and the map ΦY→Y then show that the two are DK-equivalent.

We now assume that the map $\Phi(\mathrm{sk}_{n-1}Y) \to \mathrm{sk}_{n-1}Y$ is a DK-equivalence and seek to show that the map

$$\Phi(\mathrm{sk}_n Y) \to \mathrm{sk}_n Y$$

is also for $n \geq 2$. Notice that $\mathrm{sk}_n Y$ is obtained from $\mathrm{sk}_{n-1}Y$ via iterations of pushouts of diagrams of the form

$$Q_{m,n} \longleftarrow P_{m,n} \longrightarrow \mathrm{sk}_{n-1}Y. \tag{9.2}$$

For simplicity, we will assume that $m=0$ and we require only one such pushout to obtain $\mathrm{sk}_n Y$. Notice that $(\mathrm{sk}_{n-1}Y)_0 = (\mathrm{sk}_n Y)_0 = Y_0$ and that the map

$$\mathrm{sk}_{n-1}Y \to \mathrm{sk}_n Y$$

is the identity on the discrete space in degree zero. Therefore we use the distinct n-simplex $\Delta[n]^t_{y_0\ldots\ldots y_n}$ for each $(y_0 \ldots\ldots y_n) \in Y_0^{n+1}$ as defined above in Section 3.12. Setting $\underline{y} = (y_0 \ldots\ldots y_n)$, we write this n-simplex as $\Delta[n]^t_{\underline{y}}$.

We can then apply the map Φ to diagram (9.2) (and its pushout) to obtain the diagram

$$\Phi\dot{\Delta}[n]^t_{\underline{y}} \longrightarrow \Phi\mathrm{sk}_{n-1}Y$$
$$\downarrow \qquad\qquad \downarrow$$
$$\Phi\Delta[n]^t_{\underline{y}} \longrightarrow \Phi\mathrm{sk}_n Y. \tag{9.3}$$

We would like to know that we still have a pushout diagram. In other words, we want to know that the functor Φ preserves pushouts. To see that it does, consider the level wise pullback diagram defining $(\Phi Y)_n$:

$$
\begin{array}{ccc}
(\Phi Y)_n & \longrightarrow & Y_n \\
\downarrow & & \downarrow \\
X_0^{n+1} & \longrightarrow & Y_0^{n+1}.
\end{array}
$$

We can regard the map f: $X \to Y$ as inducing a pullback functor f^* from the category of simplicial sets over Y_0^{n+1} to the category of simplicial sets over X_0^{n+1}. (Recall that the category of objects over a simplicial set Z has as objects maps of simplicial sets $W \to Z$ and as morphisms the maps of simplicial sets making the appropriate triangle commute.) However, this functor between over categories can be shown to have a right adjoint. Therefore it is a left adjoint and hence preserves push-outs.

We know that the maps

$$\Phi \dot{\Delta}[n]_{\underline{y}}^t \to \dot{\Delta}[n]_{\underline{y}}^t$$

and

$$\Phi(\mathrm{sk}_{n-1} Y) \to \mathrm{sk}_{n-1} Y$$

are DK-equivalences by our inductive hypothesis, since the nondegenerate simplices in each case are concentrated in degrees less than n. Since the left-hand vertical maps of diagrams (9.2) and (9.3) above are cofibrations, the right-hand vertical map in diagram (9.3) is also a cofibration, and therefore it remains only to show that the map $\Phi \Delta[n]_{\underline{y}}^t \to \Delta[n]_{\underline{y}}^t$ is a DK-equivalence in order to show that the push-outs of the two diagrams are weakly equivalent.

If n=0, then $\Phi \Delta[0]_{\underline{y}}^t \to \Delta[0]_{\underline{y}}^t$ is a DK-equivalence since everything is already local and $\Phi \Delta[0]_{\underline{y}}^t$ is just the nerve of some contractible category. In fact, given any $n \geq 0$ and $\underline{y} = (y_0 \ldots \ldots y_n)$, if $y_i \neq y_j$ for each

$0 \leq i, j \leq n$, the map $\Phi \Delta[n]_{\underline{y}}^t \to \Delta[n]_{\underline{y}}^t$ is a DK-equivalence, since $\Delta[n]_{\underline{y}}^t$ is already local.

Now suppose that n=1 and $\underline{y} = (y_0, y_0)$. Consider $g : \Phi \Delta[1]_{\underline{y}}^t \to \Delta[1]_{\underline{y}}^t$ and let k be the number of 0-simplices of $g^{-1}(y_0)$. If C_k denotes the category with k objects and a single isomorphism between any two objects, then we have that

$$\Phi \Delta[1]_{\underline{y}}^t \simeq \Delta[1]_{\underline{y}}^t \times \text{nerve}(C_k).$$

Thus, it suffices to show that

$$L_c \Phi \Delta[1]_{\underline{y}}^t \simeq L_c \Delta[1]_{\underline{y}}^t \times L_c \text{nerve}(C_k).$$

To prove this fact, first note that the fibrant objects in SeSp_c are closed under internal hom, namely that given a Segal space W and any simplicial space Y, there is a Segal space WY given by $(W^Y)_k = \text{Map}^h(Y \times \Delta[k]^t, W)$ [17, 7.1]. Therefore, given any Segal precategories X and Y and any Segal space W, we can work in the category SeSp_c and make the following calculation.

$$\begin{aligned} \text{Map}^h(L_c X \times L_c Y, W) &\simeq \text{Map}^h(L_c X, W^{L_c Y}) \\ &\simeq \text{Map}^h(X, W^Y) \\ &\simeq \text{Map}^h(X \times Y, W) \\ &\simeq \text{Map}^h(L_c(X \times Y), W). \end{aligned}$$

In other words, the map

$$L_c(X \times Y) \to L_c X \times L_c Y$$

is a DK-equivalence, and in particular the statement above for $L_c \Phi \Delta[n]_{\underline{y}}^t$ holds.

Now consider the case where n=2. Then if $\underline{y} = (y_0, y_1, y_2)$, we have that $G(2)^t_{\underline{y}}$ can be written as a pushout

$$
\begin{array}{ccc}
G(0)^t_{y_1} & \longrightarrow & G(1)^t_{y_0,y_1} \\
\downarrow & & \downarrow \\
G(1)^t_{y_1,y_2} & \longrightarrow & G(2)^t_{\underline{y}}.
\end{array}
$$

(9.4)

Now consider the map $g : \Phi G(2)^t_{\underline{y}} \to G(2)^t_{\underline{y}}$. We have that $g^{-1}(G(0)^t_{y_1})$ is the nerve of some contractible category. Similarly, the map $g^{-1}(G(1)^t_{y_0,y_1}) \to G(1)^t_{y_0,y_1}$ is a DK-equivalence, as is the map $g^{-1}(G(1)^t_{y_1,y_2}) \to G(1)^t_{y_1,y_2}$. Since we have a pushout diagram

$$
\begin{array}{ccc}
g^{-1}(G(0)^t_{y_1}) & \longrightarrow & g^{-1}(G(1)^t_{y_0,y_1}) \\
\downarrow & & \downarrow \\
g^{-1}(G(1)^t_{y_1,y_2}) & \longrightarrow & \Phi G(2)^t_{\underline{y}}
\end{array}
$$

and the left-hand vertical maps of this diagram and of diagram (9.4) are cofibrations, it follows that the map $\Phi G(2)^t_{\underline{y}} \to G(2)^t_{\underline{y}}$ is a DK-equivalence. In fact, for any n≥2, $G(n)^t_{\underline{y}}$ can be obtained by iterating such pushouts. Therefore, we have shown that the map $\Phi G(n)^t_{\underline{y}} \to G(n)^t_{\underline{y}}$ is a DK-equivalence.

To see that $\Phi \Delta(n)^t_{\underline{y}} \to \Delta(n)^t_{\underline{y}}$ is a DK-equivalence for any choice of $\underline{y}$, we need a variation on this argument. Again using a pushout construction, we will use the fact that this map is a DK-equivalence when each y_i is distinct to show that it is also a DK-equivalence even if $y_i = y_j$ for some i≠j. We will describe this construction for a specific example, but it works in general. Specifically, we show that $\Phi\Delta[2]^t_{y_0,y_1,y_0} \to \Delta[2]^t_{y_0,y_1,y_0}$ is a DK-equivalence.

Define the Segal precategory $\tilde{y} = \tilde{y} \amalg [\tilde{y}]$, where $\tilde{y}$ is a 0-simplex not in Y_0, and we regard

$[\tilde{y}]$ as a doubly constant simplicial space. Then, using the map g: $\Phi Y \to Y$ and some vertex y_0 of Y, we let Z be a Segal precategory isomorphic to $(g^{-1}y_0)$ and define $\tilde{X} = \amalg Z$. There is a map $\tilde{X} \to \tilde{y}$ such that Z maps to $\tilde{y}$. We define a functor $\tilde{\Phi}$ and factorization

$$\tilde{X} \longrightarrow \tilde{\Phi}\tilde{Y} \xrightarrow{g} \tilde{Y}$$

just as we defined ΦY above. More generally, we apply $\tilde{\Phi}$ to any Segal precategory with 0-simplices those of $\tilde{Y}$ to obtain a Segal precategory with 0-simplices those of $\tilde{x}$, just as we have been doing with Φ.

Now consider the objects $G(2)^t_{y_0,y_1,\tilde{y}}$ and $\Delta(2)^t_{y_0,y_1,\tilde{y}}$, each with 0-simplices those of $\tilde{y}$. There is a natural map

$$G(2)^t_{y_0,y_1,\tilde{y}} \to G(2)^t_{y_0,y_1,y_0}$$

where $\tilde{y} \mapsto y_0$, and an analogous map

$$\Delta[2]^t_{y_0,y_1,\tilde{y}} \to \Delta[2]^t_{y_0,y_1,y_0}.$$

We have a pushout diagram

$$G(2)^t_{y_0,y_1,\tilde{y}} \longrightarrow G(2)^t_{y_0,y_1,y_0}$$
$$\downarrow \qquad\qquad \downarrow$$
$$\Delta[2]^t_{y_0,y_1,\tilde{y}} \longrightarrow \Delta[2]^t_{y_0,y_1,y_0}.$$

Since the left-hand vertical map is a cofibration, this map is actually a homotopy pushout diagram.

Now, from above we know that the maps

$$\tilde{\Phi} G(2)^t_{y_0,y_1,\tilde{y}} \to G(2)^t_{y_0,y_1,\tilde{y}}$$

and

$$\Phi G(2)^t_{y_0,y_1,y_0} \to G(2)^t_{y_0,y_1,y_0}$$

are DK-equivalences. We also know that the map

$$\tilde{\Phi} \Delta[2]^t_{y_0,y_1,\tilde{y}} \to \Delta[2]^t_{y_0,y_1,\tilde{y}}$$

is a DK-equivalence since the 0-simplices $y_0, y_1, \tilde{y}$ are distinct. We can consider the pushout diagram

$$\tilde{g}^{-1} G(2)^t_{y_0,y_1,\tilde{y}} \longrightarrow g^{-1} G(2)^t_{y_0,y_1,y_0}$$
$$\downarrow \qquad\qquad \downarrow$$
$$\tilde{g}^{-1} \Delta[2]^t_{y_0,y_1,\tilde{y}} \longrightarrow \Phi \Delta[2]^t_{y_0,y_1,y_0}$$

which is again a homotopy pushout diagram. It follows that the map

$$\Phi \Delta[2]^t_{y_0,y_1,y_0} \to \Delta[2]^t_{y_0,y_1,y_0}$$

is a DK-equivalence, completing the proof.

We now proceed with the other remaining proof from Section 5.

Proof of Proposition 5.10: Suppose that f: X→Y is a fibration and a weak equivalence. First, consider the case where f_0: X_0→Y_0 is an isomorphism. Without loss of generality, assume that X_0=Y_0 and factor the map f: X→Y functorially in $SSets_c^{\Delta^{op}}$ as the composite of a cofibration and an acyclic fibration in such a way that the Y_0' remains a discrete space:

$$X \longhookrightarrow Y' \xrightarrow{\;\sim\;} Y.$$

(We can obtain a Y' with discrete zero space by taking a factorization in $SSets_c^{\Delta^{op}}$ analogous to the one we defined for $SeSp_c$ at the beginning of Section 5.) Since the map X→Y is a DK-equivalence and the map Y'→Y is a Reedy weak equivalence and therefore a DK-equivalence, it follows that the map X→Y' is a DK-equivalence. In particular, X→Y' is an acyclic cofibration and therefore by the definition of fibration in $SeCat_f$ the dotted arrow lift exists in the following solid arrow diagram:

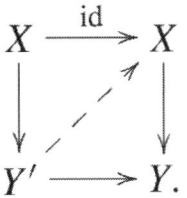

Thus, f: X→Y is a retract of Y'→Y and therefore a Reedy acyclic fibration. In particular, f has the right lifting property with respect to the maps in $I_{c'}$ since they are monomorphisms and therefore Reedy cofibrations.

Now consider the general case, where X_0→Y_0 is surjective but not necessarily an isomorphism. Then, as in the proof of Lemma 5.9, define the object ΦY and consider the composite map X→ΦY→Y. since by the first case X→ΦY has the right lifting property with respect to the maps in $I_{c'}$ it remains to show that ΦY→Y has the right lifting property with respect to the maps in I_c.

Let $A \to B$ be an acyclic cofibration. Then there is a dotted arrow lift in any solid arrow diagram of the form

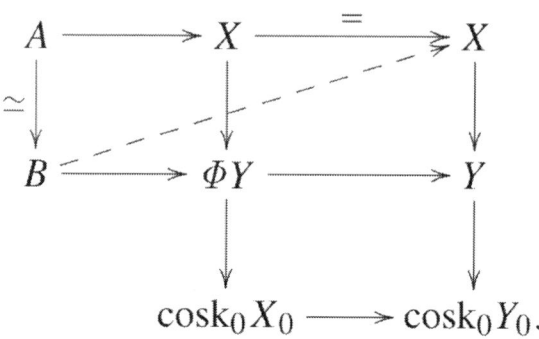

$$(9.5)$$

We would like to know that this lift $B \to X$ also makes the upper left-hand square commute.

Suppose that $A_0 = B_0 = X_0$. In this case, a map $B \to Y$ together with a lifting

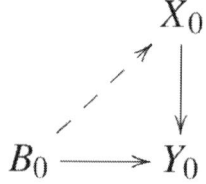

completely determines a map $B \to \Phi Y$. Therefore, in this fixed object set case, there is only one possible lifting $B \to X$ in diagram (9.5), and one which makes the upper left-hand square commute.

The map $X \to \Phi Y$ is a fibration in the fixed object model category structure $\mathrm{LSSets}_{0,f}^{\Delta^{op}}$ where $O = X_0$. However, since the cofibrations in $\mathrm{LSSets}_{0,f}^{\Delta^{op}}$ are precisely the monomorphisms, the acyclic fibrations are Reedy acyclic fibrations. Therefore, the map $X \to \Phi Y$ is a Reedy acyclic fibration and thus has the right lifting property with respect to all monomorphisms of simplicial spaces. In particular, it has the right lifting property with respect to the maps in I_c.

Using the construction of ΦY and the fact that $X \to Y$ is a fibration and a weak equivalence, we can see that $X_0 \to Y_0$ is surjective. In particular, the map $\cos k_0 X_0 \to \cos k_0 Y_0$ has the right lifting property with respect to the maps in I_c. Using the universal property of pullbacks, we can see that the map $\Phi Y \to Y$ also has the right lifting property with respect to the maps in I_c.

ACKNOWLEDGMENTS

This paper is a version of my Ph.D. thesis at the University of Notre Dame [6]. I would like to thank my advisor Bill Dwyer for his help and encouragement on this paper. I would also like to thank Charles Rezk, Bertrand Toën, and the referee for helpful comments.

The author was partially supported by a Clare Boothe Luce Foundation Graduate Fellowship.

REFERENCES

1. B. Badzioch, Algebraic theories in homotopy theory, Ann. of Math. (2) 155 (3) (2002) 895–913.
2. J.E. Bergner, A characterization of fibrant Segal categories, Proc. Amer. Math. Soc. (in press). Preprint available at math.AT/0603400.
3. J.E. Bergner, A model category structure on the category of simplicial categories, Trans. Amer. Math. Soc. 359 (2007) 2043–2058.
4. J.E. Bergner, Rigidification of algebras over multi-sorted theories, Algebr. Geom. Topol. 6 (2006) 1925–1955.
5. J.E. Bergner, Simplicial monoids and Segal categories, in: Categories in Algebra, Geometry, and Mathematical Physics, in: Contemporary Mathematics (in press). Preprint available at math.AT/0508416.
6. J.E. Bergner, Three models for the homotopy theory of homotopy theories, Ph.D. Thesis, University of Notre Dame, 2005.
7. D. Dugger, Universal homotopy theories, Adv. Math. 164 (1) (2001) 144–176.
8. W.G. Dwyer, D.M. Kan, Function complexes in homotopical algebra, Topology 19 (1980) 427–440.

9. W.G. Dwyer, D.M. Kan, Simplicial localizations of categories, J. Pure Appl. Algebra 17 (3) (1980) 267–284.
10. W.G. Dwyer, J. Spaliński, Homotopy theories and model categories, in: Handbook of Algebraic Topology, Elsevier, 1995.
11. P.G. Goerss, J.F. Jardine, Simplicial Homotopy Theory, in: Progress in Mathematics, vol. 174, Birkhauser-Verlag, Basel, ¨ 1999.
12. P.S. Hirschhorn, Model Categories and their Localizations, in: Mathematical Surveys and Monographs, vol. 99, American Mathematical Society, Providence, RI, 2003.
13. A. Hirschowitz, C. Simpson, Descente pour les n-champs. Preprint available at math.AG/9807049.
14. M. Hovey, Model Categories, in: Mathematical Surveys and Monographs, vol. 63, American Mathematical Society, Providence, RI, 1999.
15. S. MacLane, Categories for the Working Mathematician, second ed., in: Graduate Texts in Mathematics, vol. 5, SpringerVerlag, New York, 1998.
16. C.L. Reedy, Homotopy theory of model categories (unpublished). Available at http://www-math.mit.edu/˜psh.
17. C. Rezk, A model for the homotopy theory of homotopy theory, Trans. Amer. Math. Soc. 353 (3) (2001) 973–1007.
18. G. Segal, Categories and cohomology theories, Topology 13 (1974) 293–312.
19. B. Toen, G. Vezzosi, Homotopical algebraic geometry. I. Topos theory, Adv. Math. 193 (2) (2005) 257–372.

CITATION

Julia E. Bergner, Three models for the homotopy theory of homotopy theories, Topology, Volume 46, Issue 4, September 2007, Pages 397-436, ISSN 0040-9383, http://dx.doi.org/10.1016/j.top.2007.03.002.

Perspectives on A-homotopy Theory and Its Applications

Hélène Barcelo[a,]
and Reinhard Laubenbacher[b]

[a]Department of Mathematics and Statistics, Arizona State University, Tempe, AZ 85287-1804, USA
[b]Virginia Bioinformatics Institute at Virginia Tech, Washington St. (0477), Blacksburg, VA 24061, USA

ABSTRACT

This paper contains a survey of the A-theory of simplicial complexes and graphs, a combinatorial homotopy theory developed recently. The initial motivation arises from the use of simplicial complexes as models for a variety of complex systems and their dynamics. This theory diverges from classical homotopy theory in several crucial aspects. It is related to prior work in matroid theory, graph theory, and work on subspace arrangements.

INTRODUCTION

In his book [3] Atkin says: "In order to capture the geometric essence of any natural system N, we must choose an appropriate formal geometric structure into which the observables of N can be encoded. It turns out to be useful to employ what is termed a simplicial complex as our formal mathematical framework. A simplicial complex is a natural generalization of the intuitive idea of a Euclidean space, and is formed by interconnecting a number of pieces of varying dimension. The mathematical apparatus, which has its roots in algebraic topology, gives us a systematic procedure for keeping track of how the pieces fit together to generate the entire object, and how they each contribute to the geometrical representation of N."

Atkin proceeded to model a variety of social and technological networks using simplicial complexes. Examples range from soccer and its strategic subtleties to the committee structure at the University of Essex. (In the latter case simplices correspond to committees, with the members represented by the vertices. Combinatorial "holes" in the complex correspond to "missing" committees, that is, committees with a membership suitable for certain issues to be addressed.) In order to analyze and compare social structures he developed a measure on complexes which he termed Q-analysis [1] and [2]. It is reminiscent of measuring the connected components of a topological space, except that Atkin was interested in measuring the combinatorial connectivity of the complex.

The central object of Q -analysis is an integer vector associated with a simplicial complex Δ as follows. Suppose the dimension of Δ is d. Let $0 \leqslant q \leqslant d$, and let $\sigma, \tau \in \Delta$ be two simplices. Call σ and τ q-near if they share a simplex of dimension q, that is, if their intersection contains at least q+1 elements. The two simplices are q-connected if there is a sequence

$$\sigma, \sigma_1, \ldots, \sigma_n, \tau,$$

such that consecutive simplices are q-near. This notion of connectivity generates an equivalence relation on the simplices of Δ for each choice of q. Define

$$Q(\Delta) = (q_0, q_1, \ldots, q_d),$$

where q_i is the number of equivalence classes obtained by choosing q=i. Observe that for q=0 one obtains exactly the number of connected components of Δ viewed as a topological space, and for q=d one simply obtains the number of simplices of maximal dimension. Atkin and others used Q-analysis to study phenomena such as traffic flow and television viewing habits (see e.g. [20]).

Laubenbacher became interested in Q-analysis as a potential tool to analyze the dynamic network of interactions in socio-technical complex systems. One goal was to associate qualitative measures with different dynamic modes of the system. As an example, consider a collection of stock traders, say at the New York Stock Exchange. The buying and selling decisions of each individual trader depend in part on information obtained from a variety of sources, on software that analyzes market trends, and on the actions of other select traders. How is the system affected when, for instance, one or more traders are equipped with faster data links than others? As another example, consider the drug traffic interception efforts of government authorities in the Southwestern US. Through a variety of means, including blimps stationed in strategic positions along the US–Mexican border, data are collected on air and ground traffic bringing illegal drugs into Arizona, California, New Mexico, and Texas. One smuggling method is to fly drugs to clandestine air strips on the US side of the border and then use other planes and ground transport for further distribution. Is it possible to use observed air traffic patterns of a partially known network of clandestine airstrips to reconstruct the unknown part? Finally, these kinds of questions have counterparts in other systems of interactions, such as the gene regulatory network of an organism or the interaction of species in an ecosystem.

While Q-analysis is sometimes useful for questions of this sort, it is a very crude invariant of a complex, just like the set of connected components of a topological space does not contain a great deal of information about the space. Atkin had realized this and proposed a definition for a group associated with a simplicial complex, similar to the fundamental group of a pointed topological space [2]. But it too should be an invariant of certain aspect of the combinatorial rather than the topological structure of the complex. A rigorous definition of such a group was given in [22], together with an algorithm for its computation. At that point it had become clear that this group had to be part of a general theory, with Atkin›s Q-analysis representing dimension zero. The theory should be similar to the classical homotopy theory of a pointed topological space. However, it should depend on the combinatorial structure of the complex, rather than on its properties when viewed as a topological space. In applications, the individual simplices have interpretations that should not be lost in the computation of invariants. For instance, topologically, any two triangulations of a 2-sphere are equivalent, whereas combinatorially they will in general be very different.

Such a new combinatorial homotopy theory, was presented in [6], termed A-theory, in honor of Atkin. It is similar to classical homotopy theory in some respects and different in others. Similarities include such properties as a Seifert–van Kampen Theorem for the combinatorial fundamental group, a long exact sequence associated with the relative theory, and the fact that the higher dimensional groups are abelian. Differences include, for instance, the fact that complexes that are contractible as topological spaces can have nontrivial A-groups, and lack of invariance under triangulation. Using a completely different definition, a combinatorial homotopy theory $A_n^G(\Gamma)$ for graphs Γ was also defined and related to the A-theory of simplicial complexes.

$$A_n^q(\Delta, \sigma_0), \quad n \geqslant 1, \quad 0 \leqslant q \leqslant \dim(\Delta), \quad \sigma_0 \in \Delta,$$

A fascinating aspect of A-theory is that once it was well defined and applied to different simplicial complexes, it was discovered that, in fact, it is related to constructions arising in quite diverse contexts. We briefly describe three examples that will be revisited in greater depth in the last section of this paper. In the early 1970s, Maurer, in his study of matroid basis graphs, was led to develop a homotopy theory for matroid complexes (see [27, Section 4]). As Maurer mentions, the classical notion of path homotopy applies to graphs, but his notion is not the same, nor is it the same as Tutte's [30]. It turns out that the A1-group of a matroid corresponds exactly to Maurer's graph homotopy group. Later on, in 1977, Lovász [24] introduced new topological methods for proving some connectivity results in graph theory. One of his techniques consists of attaching 2-cells to all 3- and 4-cycles of a graph, before computing its (classical) fundamental group. It so happens that this computation is equivalent to calculating the A1-group of the original graph. More recently, Babson et al. [5] discovered that the An-groups of the order complexes associated with the intersection lattice of some arrangements of linear subspaces coincide with the homotopy groups of the (real) complements of those arrangements, a fact also (independently) proved by Björner [9], for the case n=1. All these constructions, and several more, can be formulated within the framework of A-theory, proving it to be an interesting theory.

In the next section we give the definition of A-theory, both for simplicial complexes and for graphs. The definitions are illustrated with examples. It is worthwhile to note that both definitions are important to understand all aspects of the theory, so both should be kept in mind. In Section 3, an algorithm for computing the abelianization of the A1-groups is described, while in Section 4, we recall some classical behavior exhibited by the A-groups, and explain how the two definitions are related. The last two sections are devoted to several applications of A-theory.

DEFINITIONS AND THEOREMS

As mentioned in the introduction, there are two frameworks for A-theory, one using simplicial complexes and the other using graphs. The two approaches are closely related and we will recall them here. All details and proofs can be found in [6].

A1 of Simplicial Complexes

We begin with a simplicial complex Δ of dimension d, a fixed integer q, with $0 \leqslant q \leqslant d$, and a given maximal simplex σ_0 (with respect to inclusion) of dimension greater than or equal to q. For further details regarding the following definitions see Section 2 of [6].

Definition 1.1

— Two simplices σ and τ of Δ are q-connected, if there is a sequence of simplices (in Δ)
$$\sigma, \sigma_1, \sigma_2, \ldots, \sigma_n, \tau,$$
such that any two consecutive ones share a q-face, that is, they have at least q+1 vertices in common. Such a chain will be called a q-chain.

— The complex Δ is q-connected, if any two simplices in Δ of dimension greater than or equal to q are q-connected.

— A q-loop in Δ based at σ_0 is a q -chain beginning and ending at σ_0. Denote a q-loop $\sigma_0, \sigma_1, \ldots, \sigma_n, \sigma_0$ by $(\sigma_0, \sigma_1, \ldots, \sigma_n, \sigma_0) = (\sigma)$. Its length is n. (Note that the σ_i need not be distinct.)

Two such combinatorial q-loops of simplices are A-homotopic if they can be deformed into each other without breaking any q-dimensional connections. More precisely, we have the following definition.

Definition 1.2

Let $\simeq A$ be the equivalence relation on the collection of q-loops in Δ, based at σ_0, generated by the following three conditions.

— The q-loop

$$(\sigma) = (\sigma_0, \ldots, \sigma_i, \sigma_{i+1}, \ldots, \sigma_n, \sigma_0)$$

is equivalent to the q-loop

$$(\sigma') = (\sigma_0, \ldots, \sigma_i, \sigma_i, \sigma_{i+1}, \ldots, \sigma_n, \sigma_0).$$

That is, loops can be "stretched" by repeating a simplex without changing its equivalence class.

— Suppose that (σ) and (τ) have the same length. They are equivalent if there is a diagram as in Fig. 1. The diagram is to be interpreted as follows. A horizontal or vertical edge between two simplices indicates that they share a q-face. Each row in the diagram is a q-loop based at σ_0, while each column represents a q -chain starting at σ_i and ending at τ_i. Thus, (σ) is equivalent to (τ) $((\sigma) \simeq A (\tau))$ if there is a sequence of q-loops based at σ_0 connecting them. Such a diagram is said to be an A-homotopy between (σ) and (τ).

— A q-loop is called A-contractible if it is A-homotopic to the constant q -loop at the base simplex σ_0.

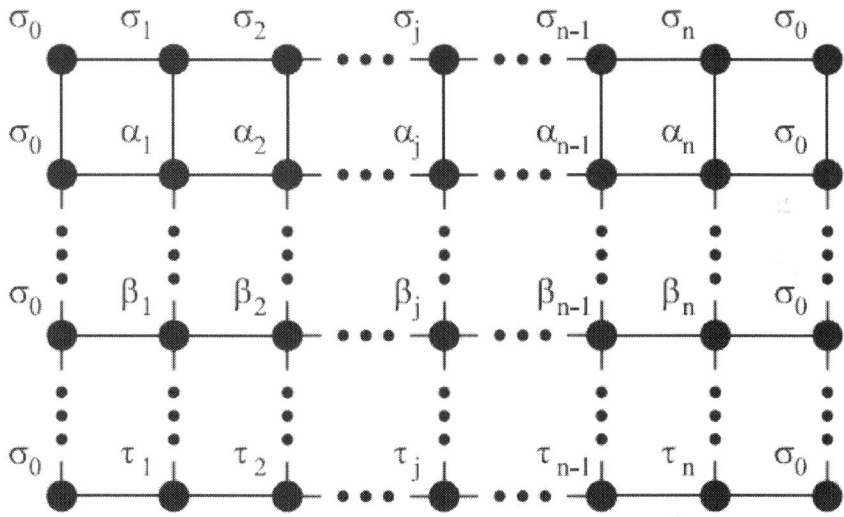

Figure 1: σ and τ, two equivalent q-loops.

This equivalence relation is called A-homotopy, and the equivalence class of a loop (σ) is denoted by$[\sigma]$, while the set of all equivalence classes is denoted by $A_1^q(\Delta,\sigma_0)$.

The next natural step is to concatenate q-loops based at σ_0 in order to obtain a product operation on $A_1^q(\Delta,\sigma_0)$. Having done so, it is easily shown that $A_1^q(\Delta,\sigma_0)$. is a group with unit element the equivalence class of the constant (or trivial) loop (σ_0). In this group, the inverse of an element $[\sigma]$ is given by the equivalence class of the same loop traversed in the opposite direction. So, we have obtained a family $A_1^q(\Delta,\sigma_0)$. of groups, one for each $0 \leqslant q \leqslant d = \dim(\Delta)$.

The subscript suggests that these definitions and groups might be extended to higher dimensions. Indeed, this is the case, and in fact, $A_1^q(\Delta,\sigma_0)$. is the A -counterpart of the fundamental group of a simplicial complex, $\pi_1 (\Delta, \sigma_0)$. The generalization of these concepts to $A_1^q(\Delta,\sigma_0)$. groups can be found in [6] and will not be reproduced here. As it turns out, they are also the A -counterpart of the higher homotopy groups of a simplicial complex, $\pi_n (\Delta, \sigma_0)$. We now describe A-theory of graphs.

A1 of Graphs

The definition of A-theory for graphs parallels closely that of the homotopy groups of a topological space. We start by recalling some elementary constructions from graph theory. For more details see Section 5 of [6].

Definition 1.3.

Let $\Gamma_1 = (V_1, E_1), \Gamma_2 = (V_2, E_2)$ be simple graphs, that is, graphs without loops and multiple edges.

— The Cartesian product $\Gamma_1 \times \Gamma_2$ is the graph with vertex set $V_1 \times V_2$. There is an edge between (u_1, u_2) and (v_1, v_2) if either $u_1 = v_1$ and $u_2 v_2 \in E_2$ or $u_2 = v_2$ and $u_1 v_1 \in E_1$.
— A graph map f: $\Gamma_1 \rightarrow \Gamma_2$ is a set map $V_1 \rightarrow V_2$ such that, if $uv \in E_1$, then either $f(u)=f(v)$ or$f(u)f(v) \in E2$.
— Let I_m be the graph with m+1 vertices labeled 0,1,...,m, and edges (i-1)i for i=1,...,m.
— Let $v_1 \in \Gamma_1$, $v_2 \in \Gamma2$ be distinguished base vertices. A based graph map f: $(\Gamma_1, v_1) \rightarrow (\Gamma_2, v_2)$ is a graph map such that $f(v_1)=v_2$.

Next, we recall G-homotopy of graph maps and G-homotopy equivalence of graphs.

Definition 1.4.

— Let f, g: $(\Gamma_1, v_1) \rightarrow (\Gamma_2, v_2)$ be based graph maps. Then f and g are called G-homotopic , denoted by $f \simeq_G g$, if there is an integer $m \geqslant 1$ and a graph map

$$\phi : \Gamma_1 \times \mathbf{I}_m \longrightarrow \Gamma_2,$$

such that $\varphi(-,0)=f$, and $\varphi(-,m)=g$, and such that $\varphi(v_1,i)=v_2$ for all i.

— We call (Γ_1, v_1) and (Γ_2, v_2) G-homotopy equivalent if there exist based graph maps f: $\Gamma_1 \rightarrow \Gamma_2$ and g: $\Gamma_2 \rightarrow \Gamma_1$ such that $gf \simeq_G \mathrm{id}\Gamma_1$ and $fg \simeq_G \mathrm{id}\Gamma_2$. The maps f and g are called G-homotopy inverses of each other.
— A graph map f: $(\Gamma_1, v_1) \rightarrow (\Gamma_2, v_2)$ is G-contractible if it is G -homotopic to the graph map that sends all vertices (thus edges) to the base vertex v_2.

The base point for the graph $\mathbf{I}_m$ will be the vertex labeled 0, and the boundary $\partial (\mathbf{I}_m)$ of $\mathbf{I}_m$ consists of the vertices labeled 0 and m. Given this, $A_1^G(\Gamma, v_0)$ is the set of G-homotopy classes of graph maps

$$f : (\mathbf{I}_m, 0) \longrightarrow (\Gamma, v_0),$$

for all $m \geqslant 1$, such that $f(\partial \mathbf{I}_m)=v_0$. Note that we allow m to vary that is, we allow arbitrarily fine subdivisions of the discrete "unit" interval, for it can be shown that two maps from the discrete unit interval of different heights can be viewed as being defined on the highest one, without change of homotopy type.

The equivalence class of a map f in $A_1^G(\Gamma, v_0)$ is denoted by [f]. For $A_1^G(\Gamma, v_0)$ to become a group, one needs an operation on its equivalence classes. Intuitively, if one represents a map $f:(\mathbf{I}_m,0) \rightarrow (\Gamma, v_0)$, by the chain $\mathbf{I}_m$ whose vertex i is labeled by f(i), for all $0 \leqslant i \leqslant m$, and where $f(0)=f(m)=v_0$, then the group operation [f]*[g] simply corresponds to "stacking" up the two labeled chains corresponding to f and g, in this order. It is a routine exercise to show that the stacking operation is well defined, and that $A_1^G(\Gamma, v_0)$ is a group.

EXAMPLES

Simplicial A-theory

— Consider the 2-dimensional simplicial complex, with four maximal faces of dimension 2, shown in Fig. 2, and let q=1.

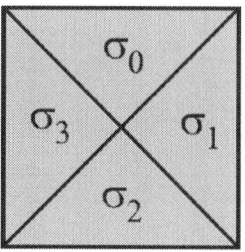

Figure 2: A 2-dimensional complex with $A_1^1 = 1$.

It is not difficult to "contract" the loop $(\sigma) = (\sigma_0, \sigma_1, \sigma_2, \sigma_3, \sigma_0)$ to the trivial constant loop (σ_0). Such a contraction is illustrated in Fig. 3. Moreover, one also easily sees that, for this complex, all the loops are A -contractible, thus making the A_1^1 group trivial. Note that the (classical) fundamental group of this complex is also trivial.

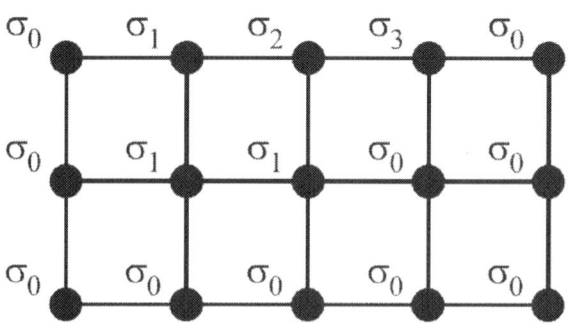

Figure 3: Contraction of the 4-loop.

1. On the other hand, if we look at the 2-dimensional simplicial complex Δ_2, shown in Fig. 4, which has five faces of dimension 2, one realizes (after some calculations) that the loop

$$(\sigma_0, \sigma_1, \sigma_2, \sigma_3, \sigma_4, \sigma_0)$$

is not A-contractible. A combinatorial explanation in terms of a "gangster problem" is given in example (5). In fact, it can be shown that the A_1^1-group for this simplicial complex is isomorphic to Z. In comparison, the (classical) fundamental group for this complex is clearly trivial, since the complex is contractible as a topological space.

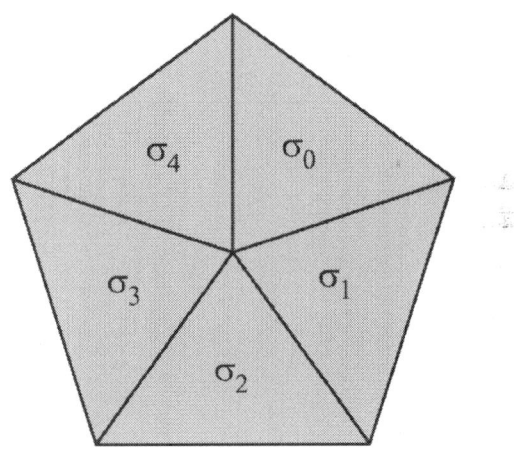

Figure 4: A 2-dimensional complex Δ_2, with $A_1^1(\Delta_2) \simeq Z$.

But there is a way to modify this complex so that the non-contractible loop becomes A-contractible. Simply "fill in the combinatorial hole" of the complex by adding a new 2-dimensional simplex as is done in Fig. 5. A contraction of the loop $(\sigma_0, \sigma_1, \sigma_2, \sigma_3, \sigma_4, \sigma_0)$ is shown in Fig. 6.

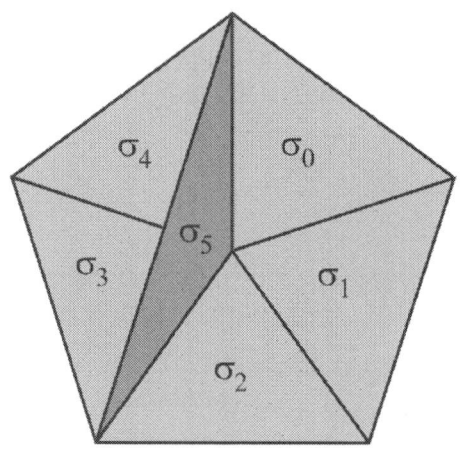

Figure 5: Filling the combinatorial hole in Δ_2. $A_1^1(\Delta_2') \simeq *$.

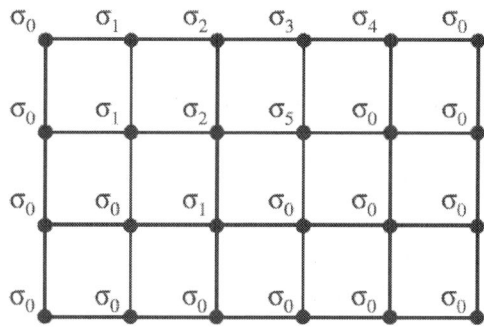

Figure 6: Contraction of the loop $(\sigma_0, \sigma_1, \sigma_2, \sigma_3, \sigma_4, \sigma_0)$.

In this case the A_1^1-group becomes trivial as is the (classical) fundamental group of this modified complex Δ_2'.

A-theory of Graphs

2. The A -theory for graphs and for simplicial complexes are very similar. Consider the graph Γ consisting of a single cycle on four vertices v_0-v_1-v_2-v_3. This cycle is G-contractible, as can be seen in Fig. 7. Indeed, a G-homotopy is given by the map

$$\phi : \Gamma \times I_2 \longrightarrow \Gamma,$$

where on $\Gamma \times \{0\}$ the map is the identity. On $\Gamma \times \{1\}$, φ is defined by $\varphi(v_0,1)=\varphi(v_3,1)=v_0$, and $\varphi(v_1,1)=\varphi(v_2,1)=v_1$. Finally, on $\Gamma \times \{2\}$ all vertices are sent to v_0. Note that for esthetic purposes, in Fig. 7 the ordered pairs $(v_{i,j})$ are labeled $v_{i,j}$. Thus the A_1^G-group of the 4-cycle graph is trivial. One also notes that there are no graph maps

$$\phi : \Gamma \times I_1 \longrightarrow \Gamma$$

that would "retract" the 4-cycle. Indeed, assuming that the interval I_m has length m=1 and that φ' is such a map, we must have $\varphi'(v_i,0)=vi$ and $\varphi'(v_i,1)=v0$ for all $0 \leqslant i \leqslant 3$. But then, φ' is not a graph map, for while $(v_2, 0)$ and $(v_2, 1)$ are adjacent in the graph $\Gamma \times I1$ their images, v_2 and v_0 (respectively), are not adjacent in Γ. One realizes that the situation in the simplicial case is analogous. That is, the 4-loop in example (1) could not be contracted to the trivial loop without going through the intermediate loop, $(\sigma_0, \sigma_1, \sigma_1, \sigma_0, \sigma_0)$. One additional remark is worth mentioning. While interpreting the 4-cycle graph Γ as a loop in R^2, one realizes that its (classical) fundamental group is nontrivial, and that in fact it is isomorphic to Z. Note, however, that attaching a 2-cell to this cycle would yield a space with (classical) fundamental group trivial, and thus equal to $A_1^G(\Gamma)$.

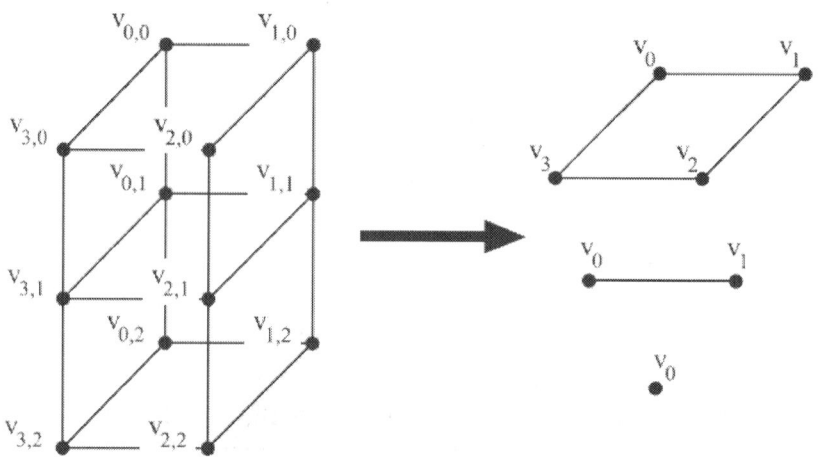

Figure 7: Contraction of the square.

3. In a similar manner, one can easily verify that the 3-cycle is G -homotopy contractible as well. In this case, the obvious map φ': Γ×I₁→Γ (similar to the one defined in example (3)) is indeed the correct graph map, which contracts the 3-cycle to a point. Again, the situation is analogous to that of the A-contraction of loops with three simplices for the simplicial approach.

4. On the other hand, the n-cycle graph, for n⩾5, does not G -contract to a point. One way of seeing it for n=5 is via the creative interpretation given by Malle [25], and known as the gangster problem:

Suppose the vertices of a graph are towns and the edges, roads connecting the towns. In each town there is a member of a gangster syndicate. The gangsters decide to meet in one of the towns. For safety reasons they decide that each day they will move from one town to an adjacent one or rest in the same town and if two of the gangsters are in adjacent towns originally, then at all steps of the journey these gangsters must be in adjacent towns, or in the same town. The problem is: For which graphs is it possible for the gangsters to meet in one of the towns?

It is not difficult to see that, indeed, the restrictions on the gangsters' movements do correspond to our notion of G -homotopy of graphs. The days represent the interval I_m (if m days are needed) and the adjacency (or resting in the same town) restriction on the movements represent the notion of graph map. Thus, drawing a 5-cycle, as in Fig. 8, with vertices labeled 1,2,3,4,5 and with the additional edge {1, 4}, one sees that the gangsters can all meet, for example, on the third day, in town 1. Indeed, on the second day, the gangsters from towns 4 and 5 moved to town 1, while the gangster from town 3 moved to town 2. On the other hand, it is clearly impossible for all the gangsters to meet at any time, in any town, if the additional edge (road) {1, 4} is not present. Again, one sees that the situation with the simplicial approach was similar. We had a non-contractible 5-loop of 2-dimensional simplices (sharing a 1-face) which could be A-contracted by filling a combinatorial hole with an additional 2-simplex.

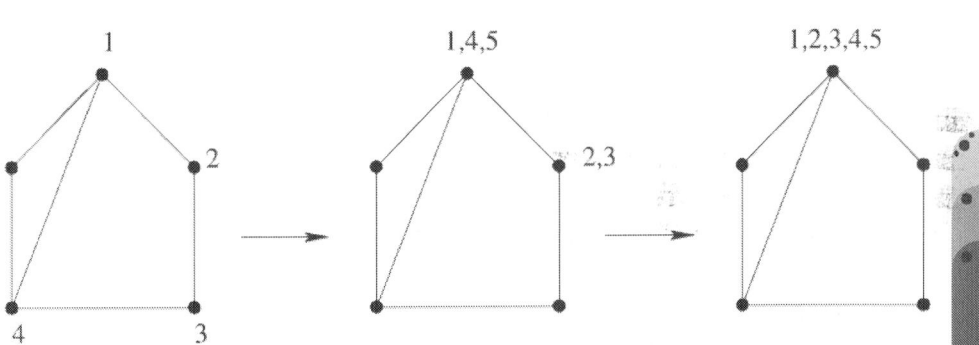

Figure 8: A gangster meeting.

CALCULATION OF $A_1^1(\Delta, \sigma_0)$

Computing the (abelianization) of A_1^q-groups turns out to be easier than one may first think. Moreover, it is via this computation that one is led to a deeper understanding of the link between the A_1^G- and A_1^q -groups. We quickly review this calculation here. For more details see

[6]. In case the reader is wondering about the base simplex (or base vertex), it should be mentioned that if σ_0 and τ_0 (or v_0, t_0) are maximal simplices in Δ (vertices in Γ) that are q-connected (connected), then

$$A_1^q(\Delta, \sigma_0) \cong A_1^q(\Delta, \tau_0), \quad (\text{or, } A_1^G(\Gamma, v_0) \cong A_1^G(\Gamma, t_0)).$$

Let $\Gamma = \Gamma^q(\Delta)$ be the graph with vertices corresponding to all simplices of Δ of dimension greater than or equal to q. Two vertices v and w are connected by an edge if and only if the corresponding simplices σ and τ share a q-face. Let v_0 be the distinguished vertex of Γ, corresponding to σ_0. This graph is said to be the q-connectivity graph of Δ. One realizes that there is a one-to-one correspondence between q-loops in Δ based at σ_0 and cycles in Γ that contain v_0. Recall that the topological fundamental group $\pi_1(\Gamma, v_0)$ is a free group with free generators. Moreover, to each cycle of Γ one can associate a specific element of $\pi_1(\Gamma, v_0)$. Let N be the normal subgroup of $\pi_1(\Gamma, v_0)$ generated by the elements corresponding to the 3- and 4-cycles of Γ.

Theorem 3.1 Barcelo et al. [6, Theorem 2.7]
$$A_1^q(\Delta, \sigma_0) \cong \pi_1(\Gamma, v_0)/N.$$

It is also worth mentioning that one can replace Γ^q by the generally much smaller graph, Γ_{max}^q, whose vertices correspond to all maximal simplices (with respect to inclusion) of Δ of dimension greater than or equal to q. From this theorem it is not too difficult to understand why the following one holds true.

Theorem 3.2 Barcelo et al. [6, Theorem 5.16]
Let Δ be a simplicial complex, with distinguished maximal simplex σ_0, $0 \leqslant q \leqslant \dim(\Delta)$. Let $\Gamma^q(\Delta)$ be the connectivity graph of Δ in dimension q, and $\Gamma_{max}^q(\Delta) \subseteq \Gamma^q(\Delta)$ be the subgraph as defined above. Then

$$A_1^q(\Delta, \sigma_0) \cong A_1^G(\Gamma^q(\Delta), v_0) \cong A_1^G(\Gamma_{max}^q(\Delta), v_0).$$

Now that the relation between the A_1^G and A_1^q groups is well established, we can give a unified definition for the higher A-homotopy groups. Let

$$\mathbf{I}_m^n = \mathbf{I}_m \times \cdots \times \mathbf{I}_m$$

denote the n-fold Cartesian product of $\mathbf{I}_m$ for some m. $\mathbf{I}_m^n$ is called an n-cube of height m. Its distinguished base point is O= (0,..., 0), and its boundary, $\partial \mathbf{I}_m^n$, is the subgraph of $\mathbf{I}_m^n$ containing all vertices with at least one coordinate equal to 0 or m. This being said, one can show that the subscript m in the above notation can be "omitted".

Definition 3.3

Let $A_n^G(\Gamma, v_0), n \geqslant 1$, be the set of homotopy classes of graph maps

$$f : (\mathbf{I}^n, \mathbf{O}) \longrightarrow (\Gamma, v_0),$$

such that $f(\partial \mathbf{I}^n) = v_0$. For n=0, $A_0^G(\Gamma, v_0)$ is the pointed set of connected components of Γ, with the component containing v_0 as distinguished element. The equivalence class of a map f in $A_n^G(\Gamma, v_0)$ is denoted by [f].

Since all the boundary points of an n-cube are given the value v_0, one easily sees that the operation of "stacking" cubes makes sense (for $n \geqslant 1$). As one expects, it can be shown that the sets $A_n^G(\Gamma, v_0)(n \geqslant 1)$ are groups, and that Theorem 3.2 holds true for all $n \geqslant 1$.

So, if one computes the fundamental group of a graph, $\pi_1(\Gamma, v_0)$, and quotients out the normal subgroup generated by all 3- and 4-cycles, one obtains the A1-group of this graph. But, inspired by Lovász' technique introduced in [24], one sees that the fundamental A_1^G-group of a graph Γ is isomorphic to the classical fundamental group of the topological space X_Γ obtained from Γ by attaching 2-cells along the boundary of each 3- and 4-cycle of Γ:

$$A_1^G(\Gamma, v_0) \cong \pi_1(X_\Gamma, v_0).$$

This was an important milestone enabling one to connect the A -theory to a wide variety of situations. Of course, the next natural question is whether there is an analogous topological space that can be constructed for all n⩾2. The answer is yes, but requires more detail and (hard) work than one would initially envision. The details can be found in [4]. Intuitively, the space X_r is a cell complex obtained by successively attaching (for m=1, 2,...) m-dimensional cells to the m-cubes of Γ (possibly degenerate m-cubes; this is analogous to the fact that the 3-cycle (triangle) can be viewed as a degenerate 4-cycle (square)), yielding a cubical complex, that bears some resemblance to Kan complexes. By this we (roughly speaking) mean that if all the faces of an m-cube belong to the space X_r then the m-cube itself belongs to the space. This is another important step, for it connects A-theory to the realm of (real) arrangements of linear subspaces.

CLASSICAL BEHAVIOR

As we saw through the examples of Section 2, A-theory and classical homotopy theory behave quite differently at times. It is now time to explore their similarities. As we saw in the last section, the $A_n^q(\Delta, \sigma_0)$ and $A_n^G(\Gamma, v_0)$ groups are intimately related. Thus, we shall refer to both of them at once, using An-groups for notation.

One of the first similarities comes from the fact that the A1-fundamental groups, like their π_1 counterparts, are not abelian in general. Moreover, for all n⩾2, the A_n-groups are abelian, likewise for the π_n groups of topological spaces.

Another common characteristic is a Seifert–van Kampen theorem. That is, one may ask whether one can compute A_1 of a simplicial complex (graph) as the free product of A_1 of appropriate subcomplexes (subgraphs), modulo A_1 of their intersection. This is indeed true, even though one must add an extra condition on the intersection of the subcomplexes (subgraphs), in addition to requiring q-connectivity of the complex (graph), the subcomplexes (subgraphs), and the intersection. This extra condition amounts to requiring that if there are 3- or 4-cycles with some of their simplices (vertices) belonging to the intersection then these 3- and 4-cycles must entirely lie in one of the subcomplexes (subgraphs) that are being intersected.

The last similarity we shall mention here is in connection to a relative A-theory. Relative A-groups with respect to a subcomplex (subgraph) have been defined. Briefly, given a subgraph (subcomplex) $\Gamma' \subset \Gamma$, and a distinguished (n-1)-face F of I^n, $(n \geqslant 2)$, the relative A_n (Γ, Γ', v_0) group is the set of all A-homotopy classes of graph maps $f:(I^n,O) \rightarrow (\Gamma,v_0)$,

such that F is mapped into Γ', together with the natural multiplication. One familiar with the classical homotopy theory will recognize this definition as the A -analog of the relative homotopy theory for simplicial complexes. So we do obtain a long exact sequence, and, as expected, the A_n-relative groups are abelian.

APPLICATIONS

In this section we review recent applications of A-theory and connections to work related to A-theory.

Maurer's Approach

In [27] Maurer studied matroid basis graphs. The basis graph of a matroid has a vertex for each basis and an edge for each pair of bases that differ by the exchange of a single pair of elements. It was well known that, for any connected graph Γ, the spanning trees, viewed as sets of edges, form the bases of a matroid. The basis graphs, called tree graphs (for the vertices correspond to the spanning trees of Γ and are denoted here by Γ_T), of such matroids had been already extensively studied [15], [17], [19] and [28]. Maurer's contribution was to completely characterize basis graphs for all matroids. For this, he studied the common neighbor (CN) subgraphs of the basis graph of a matroid.

In a graph G, if the distance between two vertices v, v' is 2, then the set of vertices consisting of v, v' and all vertices adjacent to both is called a common neighbor subgraph. Maurer's main theorem (in its first form) is the following:

Theorem 5.1 Maurer [27, Theorem 2.1]
G is a basis graph of a matroid if and only if:

— it is connected;
— each common neighbor subgraph is a square, a pyramid, or an octahedron;
— in every leveling each common neighbor subgraph satisfies the positioning condition; and
— for some v_0 the neighborhood subgraph $N(v_0)$ is the line graph of a bipartite graph.

We shall not go into details of conditions (3) and (4). Suffice it to say that while the first two conditions are relatively easy to verify, these other two are generally not reasonably dealt with. In fact, this is the reason that motivated Maurer to look for a condition that might replace them. For this, he developed the following notion of homotopy.

1. If the distance between two vertices v_{k-1} and v_{k+1} of G is equal to 2, then the paths $P_1 = v_1 \cdots v_{k-1} v_k v_{k+1} \cdots v_n$ and $P_2 = v_1 \cdots v_{k-1} v'_k v_{k+1} \cdots v_n$ are said to differ by a 2-switch. In our language this simply means that the 4-cycle $(v_k - 1 v_k v_k + 1 v'_k)$ allows us to A_1-deform P_1 into P_2 as seen in Fig. 3.

2. If the distance between v_{k-1} and v_{k+1} is equal to 1 and $P_3 = v_1 \cdots v_{k-1} v_{k+1} \cdots v_{n'}$ then P_1 and P_3 are said to differ by a shortcut. For us, this means that the 3-cycle $v_{k-1} v_k v_{k+1}$ allows us to directly A_1-deform P_1 into P_2.

3. If $v_{k-1} = v_{k+1}$ and $P_4 = v_1 \cdots v_{k-1} v_{k+2} \cdots v_{n'}$ then P_1 and P_4 are said to differ by a deletion. Again, for us this simply means that the path P_1 can be A_1-deformed into the path P_4 by simply repeating the vertex v_{k-1}.

4. Finally, Maurer declares two paths homotopic if one can be transformed into the other by a finite sequence of these elementary deformations.

It is easy to see that Maurer's notion of homotopy is equivalent to our notion of A_1-homotopy. Then Maurer goes on to prove that if Γ is a basis graph, then any two paths with the same end-points are homotopic. This is simply stating that $A_1^G(\Gamma)$ is trivial. A word of caution is in order here. While it is tempting to deduce that "surely" the A_1^G-group of a graph whose CN subgraphs are either a square, a pyramid or an octahedron must be trivial (since these subgraphs are clearly A-contractible, as they consist of 3- and 4-cycles), this is not necessarily the case. One look at Maurer's proof reveals that conditions (3) and (4) are needed to prove this theorem.

But the interesting fact is that Maurer hoped that his notion of homotopy would replace conditions (3) and (4) of his main theorem. Indeed, Maurer finishes his paper with the conjecture (still open) that Γ is a basis graph if and only if

1. Γ is connected;
2. each CN is a square, pyramid, or octahedron; and
3. $A_1^G(\Gamma)$ is trivial.

One last important remark is that Maurer also claims (without proof) that his notion of homotopy (thus our notion of A_1^G) might be the right one for graphs since

$$A_1^G(\Gamma_1 \times \Gamma_2) = A_1^G(\Gamma_1) \times A_1^G(\Gamma_2),$$

where $(\Gamma_1 \times \Gamma_2)$ denotes the Cartesian product of graphs, and $A_1^G(\Gamma_1) \times A_1^G(\Gamma_2)$ represents the direct product of groups. This is not difficult to show using our definition of A_1^G.

Lovász's Approach

In 1975, at the Fifth British Combinatorial Conference in Aberdeen, Frank [18] and Maurer [26] presented the following problem.

Theorem 5.2

Let Γ be a k-connected graph, $\{v_1,\ldots,v_k\}\subseteq V(\Gamma)$, and $n_1,\ldots,n_k$ positive integers with $n_1+\cdots+n_k=n=|V(\Gamma)|$. Then there exists a partition $\{V_1,\ldots,V_k\}$ of $V(\Gamma)$ such that

(1) $v_i\in V_i$,
(2) $|V_i|=n_i$,
(3) V_i spans a connected subgraph of Γ (i=1,...,k).

The case k=2 is rather easy, and Frank, Milliken, Györi and Lovász independently provided solutions for the case k=3. But the most interesting one was provided by Lovász, for it was the only solution that could be generalized to all $3\leqslant k\leqslant n$. The idea of the proof was based on the following innovative concept.

Given a graph Γ, add a new point a and connect it to $v_1,\ v_2,\ldots,v_k$. Denote this new graph by Γ'. Next, construct a graph $\tilde{\Gamma}$ whose vertices are the spanning trees T_i of Γ', and whose edges $(T_i,\ T_j)$ are pairs of spanning trees whose intersection $T_i\cap T_j$ contains a tree on n-1 vertices including a. So T_j can be obtained from T_i by replacing an end line by another end line. Then Lovász proved that if Γ is k -connected, then $\tilde{\Gamma}$ is connected. Finally, the last step in the proof consisted in constructing a cellular complex C_Γ (called the arborescence complex of Γ, relative to a) which is

- Simply connected and
- For which the homology groups (relative to a) $H^0(C_\Gamma)=\cdots=H^{k-2}(C_\Gamma)$ are all trivial, whenever Γ is k-connected.

One notices how similar the construction of $\tilde{\Gamma}$ is to the construction of the tree graph Γ_T associated to Γ (as described in the subsection on Maurer›s work). In fact, this connection is further developed and studied by Björner et al. in [10]. As Björner mentions [8], in the language of greedoids (which matroids are), Lovász›s arborescence complex is the basis complex (in Maurer›s sense) of the branching greedoid determined by the rooted graph Γ'.

Moreover, also notice the strong similarity with our concept of q-connectivity. We build graphs from simplicial complexes by setting the vertices to be q-simplices and edges between two of them if they are sufficiently (q-) connected.

But the similarity does not stop here. While we do not want to recall the construction of the full arborescence complex C_r, it is instructive to recall it in the case k=3. Consider the triangles (3-cycles) and quadrilaterals (4-cycles) in $\tilde{\Gamma}$ and span a 2-cell on each of them, to get the topological space C_r. On sees immediately that this is also the space $X_{\tilde{\Gamma}}$ described in Section 3. Lovász goes on proving that if Γ is a 3-connected (k-connected as well) graph then the (classical) fundamental group of the cell complex C_r is trivial, yielding (for the case k=3) in the A-language, that the A_1-group of $X_{\tilde{\Gamma}}$ is trivial.

The fact that this group is trivial is the last ingredient enabling Lovász to prove Theorem 5.2. Even though we have not yet worked out all the details, we do believe that Lovász's arborescence complex would correspond to a subcomplex of our general infinite dimensional cell complex X_r, and that the property that the fundamental group of the arborescence complex is trivial would also translate to the A-group of the appropriate subcomplex of X_r being trivial, thus enlarging the pool of applications of A-theory.

Malle's Approach

In 1983, Malle [25] developed a homotopy group for graphs, and defined what he calls the string group S (Γ) of a graph. It turns out that $S(\Gamma) = A_1^G(\Gamma)$. While Malle realized that the (classical) fundamental group of a graph is isomorphic to his string group, S (Γ), whenever a graph has girth greater than or equal to 5, he does not make the step of showing that S (Γ) is indeed isomorphic to the quotient $\pi_1(\Gamma)/N$ where N is the normal subgroup of $\pi_1(\Gamma)$ generated by 3- and 4-cycles. He also does not generalize his notion of string groups to higher dimensions.

On the other hand, Malle gives a complete description of graphs that have trivial A_1^G-group.

Theorem 5.3 Malle [25, Theorem 6]

A graph Γ has trivial A_1^G-group if and only if it is connected and each cycle of Γ has a pseudo planar net in Γ.

While we will not give the details of the definitions of planar and pseudoplanar nets, a look at Fig. 9 reveals how it works. From the planar net (left) one sees that the lower cycle is easily A-deformed into the upper one via a series of 4-cycles, then the upper cycle is A-contracted to a point via 3-cycles. In the pseudoplanar net (right), the situation is a bit different: the lower cycle is A-deformed (via 4-cycles) into two 6-cycles which in turn are both A-contracted to a point.

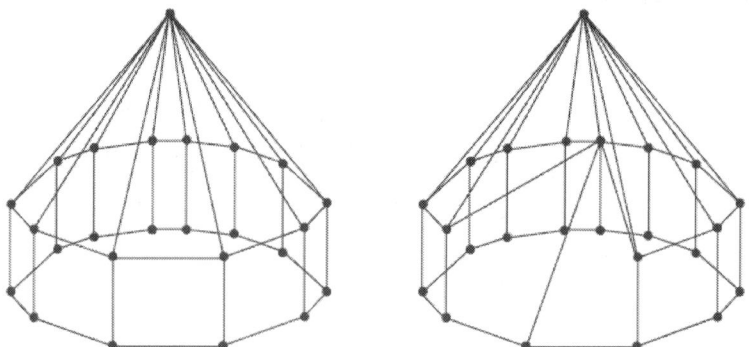

Figure 9: Planar (left) and pseudoplanar (right) nets.

Link to Subspace Arrangements

After having developed A-theory it is natural to try it on diverse simplicial complexes, in particular those interesting to combinatorialists. The first author admits to being entirely biased toward the order complexes. Given a poset P, its order complex $\Delta(P)$ is the simplicial complex on the vertex set P whose k-faces are the k-chains $x_0 < x_1 < \cdots < x_k$ in P.

The first poset for which we computed (the abelianization) of A_1^q -groups was the Boolean lattice B_n, that is, the poset of all subsets of the set $\{1, 2,...,n\}$, ordered by inclusion. The first computations gave the following results. (They were carried out with software written by Luis Garcia, available from the authors.)

— $A_1^0(\Delta(B_3))^{ab}$ is a free abelian group on 1 generator,

— $A_1^1(\Delta(B_4))^{ab}$ is a free abelian group on 7 generators,

— $A_1^2(\Delta(B_5))^{ab}$ is a free abelian group on 31 generators,

— $A_1^3(\Delta(B_6))^{ab}$ is a free abelian group on 111 generators,

— $A_1^4(\Delta(B_7))^{ab}$ is a free abelian group on? Generators?

For a long time the orders of the next groups, $A_1^q(\Delta(B_n))^{ab}$ (for $n \geqslant 7$ and $q=n-3$), remained unknown, for the computational complexity associated with the construction of the order complex of a lattice grows very rapidly. No closed formula was known nor even conjectured despite attempts (prior to 2001) at finding such a sequence in the On-Line Encyclopedia of Integer Sequences by Sloane [29]. What was then known is the following: To compute $A_1^{n-3}(\Delta(B_n))$ one draws the graph $\Gamma_{max}^{n-3}(\Delta(B_n))$ whose vertices correspond to the maximal chains of B_n, and whose edges correspond to pairs of maximal chains that differ in exactly one place. But one realizes that this coincides with the 1-skeleton of the permutahedron. Indeed, the permutahedron Π_{n-1} is defined as the convex hull of all vectors that are obtained by permuting the coordinates of the vector $(1, 2,...,n)$. Its vertices can be identified with the permutations of S_n (the symmetric group over n elements) in such a way that two vertices are connected by an edge if and only if the corresponding permutations differ by an adjacent transposition. The permutahedron is a classical object; see [31] for a nice account of its combinatorial properties. One notes that there are no 3-cycles in the 1-skeleton of the permutahedron, for this would mean that one could write the identity permutation as a product of three transpositions; the same reasoning yields that in fact there are only even length cycles. A picture of the 3-dimensional permutahedron Π_3, that is, the convex hull of all the vectors obtained by permuting the coordinates of the vector (1,2,3,4) can be seen in Fig. 10.

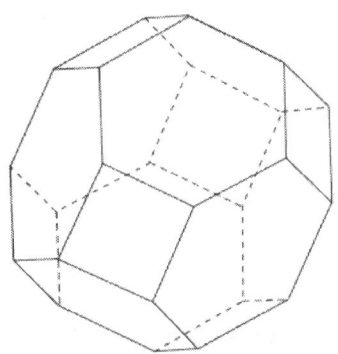

Figure 10: The 3-dimensional permutahedron, Π_3.

Thus, computing (the abelianization of) $A_1^{n-3}(\Delta(B_n))$ corresponds to computing the fundamental group of the 1-skeleton of the permutahedron to which one attaches 2-cells to each 4-cycle. For example, in the case of Π_3 after filling the 6 squares, one is left with 8 hexagons, of which 7 are generators for $A_1^1(\Delta(B_4))$ Even though one had a nice description of the $X\Gamma(\Delta(B_n))$-space no one knew how to compute its homotopy group. That is, until January 2001, when Eric Babson realized that this construction (adding 2-cells to each 4-cycle of the 1-skeleton of the permutahedron) yields a space which is homotopy equivalent to the complement $M_{n,3}$ of the (real) 3-equal arrangement!

The space $M_{n,3}$ is defined as follows: For $2 \leqslant k \leqslant n$, let $V_{n,k}$ be the set of points $\mathbf{x}=(x_1,x_2,...,x_n) \in \mathbf{R}_n$ such that $xi_1 = xi_2 = \cdots = xi_k$ for some k-set of indices $1 \leqslant i_1 < i_2 < \cdots < i_k \leqslant n$. $M_{n,k} = \mathbf{R}_n - V_{n,k}$. The k-equal arrangements have been extensively studied, and, for example, Björner and Welker [13] had a formula for the dimensions of the corresponding homology groups. Thus Babson was able to conclude that $A_1^{n-3}(\Delta(B_n))^{ab}$ is a free abelian group on generators.

$$2^{n-3}(n^2 - 5n + 8) - 1$$

In the meantime the "Bjorner–Welker" sequence of integers had also been entered into the Sloane collection of integer sequences, since it occurs in other contexts as well.

In the language of A-theory, Babson's result is as follows:

Theorem 5.4 Babson [5]

$$A_1^{n-3}(\Delta(B_n)) \cong \pi_1(M_{n,3}).$$

Proof:

We shall give a very informal proof here, since the ideas are simple while the necessary notation (and formal concepts) would only obscure the argument.

First, recall that the braid arrangement consists of all the (real) hyperplanes $H_{i,j} = \{(x_1,\ldots,x_n) | x_i = x_j\}$, for $1 \leqslant i < j \leqslant n$. The 3-equal (k-equal) arrangement embeds in the braid arrangement as each subspace of $V_{n,3}$ ($V_{n,k}$) is an intersection of some of its hyperplanes. Given a (finite) hyperplane arrangement in $\mathbf{R}^n$, its intersection with the (real) (n-1)-dimensional sphere, S^{n-1} yields a cell complex. The dual of this cell complex is a zonotope (i.e., the Minkowski sum of the line segments which are normal to the hyperplanes in the arrangement). It turns out that for the braid arrangement the dual zonotope is the permutahedron.

Next, it is also known that the complement in $\mathbf{R}^n$ of the subspace arrangement is homotopic to the space obtained by removing the faces of the zonotope corresponding to the subspaces belonging to the arrangement. For a proof of this fact see Proposition 3.1 in [14]. Let us see what those faces are for the 3-equal arrangement when n=4. First, one must remove the interior of the permutahedron (the unique 3-dimensional face), since it corresponds to the subspace of all vectors of the form (x, x, x, x), for $x \in \mathbf{R}$. Second, one must remove all 2-dimensional hexagons since they correspond to subspaces similar to $\{(x, x, x, y) | x, y \in \mathbf{R}\}$, that is, a set of vectors with three of their coordinates equal. One does not remove the 2-dimensional squares, for those correspond to subspaces similar to $\{(x, x, y, y) | x, y \in \mathbf{R}\}$ which do not belong to the

3-equal arrangement. One then realizes that this space is certainly homotopic to the one obtained by attaching 2-cells to the squares of the 1-skeleton of the permutahedron Π_3.

With a bit more effort one can see how this argument generalizes to the following theorem, also independently proved by Björner [9].

Theorem 5.5 Babson [5]

$$A_m^q(\Delta(B_n)) \cong \pi_m(M_{n,n-q}).$$

We should remark that the interest in obtaining information on these spaces $M_{n,k}$ arose in connection with a problem from computer science (see [11] and [12]). It was shown that the Betti numbers of $M_{n,k}$ are the essential ingredient in finding a lower bound for the complexity of deciding membership in $V_{n,k}$, using linear decision trees. More precisely the link with the k-equal arrangement comes from the k-equal problem: given n real numbers $x_1, x_2, \ldots, x_n$, and an integer $k \geqslant 2$, how many comparisons $x_i \geqslant x_j$ are needed to decide if some k of them are equal? Note that the comparisons $x_i - x_j \geqslant 0$ are special cases of the so-called linear tests: $l(x) \geqslant 0$. Thus the geometric reformulation reads:

Given the subspace arrangement $V_{n,k}$, how many linear tests are needed (by the best algorithm in the worst case) to decide if $x \in V_{n,k}$ for points $x \in \mathbf{R}^n$?

Thus it was natural to expect that the topological complexity of the arrangement had some bearing on the complexity of the algorithm. This is where the Betti numbers of $M_{n,k}$ entered the scene. For more details on this topic see [7], [11] and [12].

Link to Pseudomanifolds

There is one last connection to yet another area of mathematics that deserves to be mentioned. Recently, Joswig [21] introduced a (finite) group of projectivities, $\Pi(\Delta)$, for each simplicial complex Δ that is strongly connected, finite dimensional, and pure. The motivation to introduce such groups was to solve a coloring problem for simplicial polytopes which arose in the area of toric algebraic varieties. Joswig's idea consists in associating a finite group to each facet (maximal face with respect to inclusion) of Δ. For this, he first constructs what he calls the dual graph, $\Gamma(\Delta)$, of Δ. The vertices of this graph are the facets of Δ and there is an edge between two facets if the two facets share a codimension 1 face! This is exactly our $\Gamma_{max}^{d-1}(\Delta)$ if the dimension of Δ is equal to d. Even though Joswig works with pure simplicial complexes (all facets have the same dimension) his definition of the dual graph is valid for non-pure complexes as well. Next, for each codimension 1 face contained in two facets σ, τ there is a unique element $v(\sigma, \tau)$ which is contained in σ but not in τ. Joswig's perspectivity $[\sigma, \tau]:\sigma \to \tau$ is defined by setting

$$
w \mapsto \begin{cases} v(\tau, \sigma) & \text{if } w = v(\sigma, \tau), \\ w & \text{otherwise.} \end{cases}
$$

(5.1)

For a path $g = (\sigma_0, \sigma_1, \ldots, \sigma_n)$ in $\Gamma_{max}^{d-1}(\Delta)$ the projectivity $[g]$ (from σ_0 to σ_n) along g is the concatenation of perspectivities.

$$[g] = [\sigma_0, \sigma_1, \ldots, \sigma_n] = [\sigma_0, \sigma_1][\sigma_1, \sigma_2][\sigma_2, \sigma_3] \cdots [\sigma_{n-1}, \sigma_n]$$

Thus the map $[g]$ is a bijection from σ_0 to σ_n. As usual, a loop based at σ_0 is simply such a path that starts and ends at σ_0. Joswig then realizes that the projectivity of the concatenation of two paths is simply the concatenation of the corresponding projectivities, that is, $[g*h] = [g][h]$. Then Joswig's group of projectivities of Δ at σ_0, $\Pi(\Delta, \sigma_0)$, is the set of projectivities along loops based at σ_0. It is a (permutation) subgroup of the symmetric group on the set of vertices of σ_0. Despite the strong similarity with the $A_1^{d-1}(\Delta, \sigma_0)$ the groups are not isomorphic in general. Nevertheless, both groups have similar properties. Moreover, one of the connections between these groups was discovered by De Longueville and Reiner [23]. They showed that if Δ is a d-simplicial pseudomanifold which does not contain a triangle as a minor (i.e., a link in some vertex-induced subcomplex) then there is a well-defined surjective homomorphism from $A_1^{d-1}(\Delta, \sigma_0)$, dealing with "galleries of facets" of Δ, to Joswig>s group of projectivities Π (Δ, σ_0).

There are still several other applications of A-theory to other branches of mathematics such as thewonderful models of subspace arrangements, as introduced by De Concini and Procesi in [16], and buildings, to name a few. We close this paper with an application to the type of problems that provided the initial motivation for the development of A-theory.

COMBINATORIAL TIME SERIES ANALYSIS

In this example we use A-theory to analyze multivariate time series of data in cases where additional information about the local correlation between variables is available. This is the case, for instance, when the time series arises from agent-based computer simulations and we have knowledge about the interaction of agents at each time step. The techniques developed are applicable to a wide range of real and simulated systems, including such diverse examples as search-and-rescue operations and router networks for Internet packet traffic. Both examples can be represented as autonomous agent systems with a need for coordination through information exchange. Individual agents make decisions by interacting with other agents, and coordination, or control, of the system relies in an essential way on an understanding of the global structure of interaction flow.

We associate with a time series of system data a partially ordered set from which in turn we derive a simplicial complex that provides global models for the dynamic structure generated by local variable interdependence. The feature of special interest to us are the structural properties of the flow of interactions in such interaction networks, ways to measure and characterize it, and, ultimately, the ramifications of these measures for a control theory of interaction networks.

As an example we discuss here a computer simulation that the second author has co-developed, together with Michael Coombs at New Mexico State University's Physical Science Laboratory, and Abdul Jarrah, presently at VBI. First, we briefly describe the simulation, which we call AGENT, in its simplest form. AGENT consists of autonomous agents $x_1,\dots,x_n$, each of which performs an unspecified task that takes m time steps. After completing the task, each agent follows a procedure whereby it must file a report in an external database before being allowed to continue. The database, however, can only process a limited number of agents at a time (has a limited channel capacity set by the parameter S). Thus, when the number of agents ready to report exceeds the size of the channel, a queue of waiting agents forms. The size of the queue at any one time will depend on both the size of the channel and the degree of agent synchronization (i.e., how many agents are in need of the database at any one time). An agent's goal is to maximize the percentage of time that it is working at its task (i.e., it is assumed to be delay adverse). Its "fitness" is, therefore, defined as the percentage of elapsed time that it is not delayed. Agents, therefore, have an interest in desynchronizing from those with whom they are frequently in contention for access to the database. Such desynchronization is regulated by a system-wide protocol whereby, having experienced a certain degree of contention, an agent can impose desynchronizing delays on the agent, or agents, who have blocked its database access. The research described here concerns the analysis of patterns of desynchronization interactions arising from agents following this protocol.

The state of agent x_i is a vector

$$w_i = (p_1, \dots, p_{i-1}, p_{i+1}, \dots, p_n),$$

where p_j is the number of times that agent x_i was delayed by agent x_j. Each agent x_i has assigned to it a threshold T_i which represents its tolerance to delay. When $\Sigma_{jpj} > T_i$, then agent x_i selects a fixed number P_i of agents and delays them one time step (i.e., the response multiplier—"fanout"—parameter). The agents to be delayed are selected by x_i using a decision function fi attached to x_i. This function can be selected to be either deterministic or stochastic. After delaying the other agents, x_i changes its state w_i to the zero vector. The system is initialized by assigning vectors w_i to all agents. For the simulation results described here, the parameters were set to be the same for all agents, that is, $P_i=P$ and $T_i=T$ for all i. The channel capacity parameter S is set as constant for a simulation run.

For each simulation of m time steps we construct a poset as follows. It has elements x_{ij}, i=1,...,n, j=1,...,m, corresponding to agents x_i at time j. It is best to think of these as arranged in horizontal rows, with j indicating the row number. The order relation is generated by the following (covering) rule:

$$x_{ij} < x_{k,j+1},$$

if agent x_i delays agent x_k during the transition from time j to time j+1. Associated with this poset, or segments of it, we can now consider two simplicial complexes, its order complex and its covering complex, which is the simplicial complex generated by the lower-order ideals. For purposes of computation the order complex is of limited value, since it is generally far too large. The findings below made use of the covering complex only.

It was observed that the AGENT synchronization protocol yields many different mechanisms both across parameterizations, and within a single pair of parameter settings. The invariants of the simplicial complexes provide such a rich range of options for defining structure in collective desynchronization events within a single time step that the cataloging of mechanisms has only just begun. However, as an example of what there is to discover, we describe one mechanism. This involves a very interesting phenomenon we have termed "dynamic clustering". Let Δ be the covering complex for the interaction poset of AGENT for a specified time series, and fix q. A dynamic q-cluster of AGENT is a complete sub graph of the Γ^q-graph of Δ. Fig. 11 contains an example of a 3-time-step output from AGENT. Fig. 12 illustrates the dynamic q-clusters for each time step.

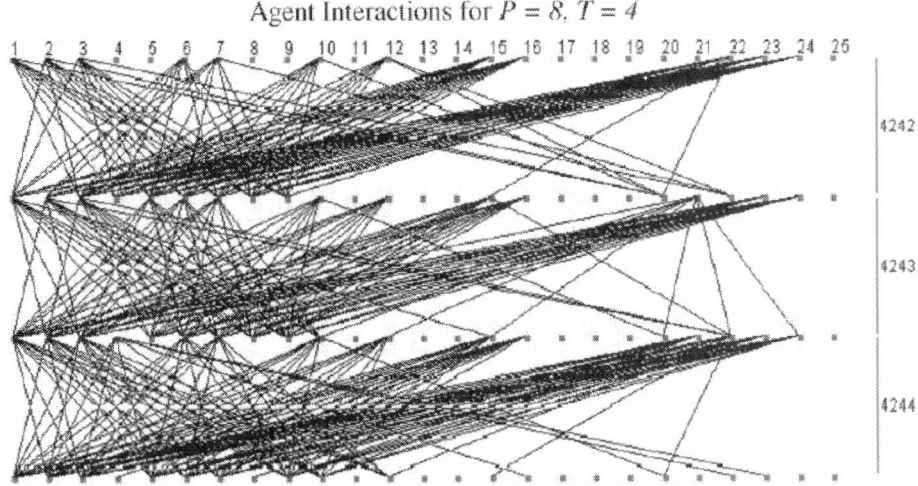

Figure 11: Interaction poset over three time steps.

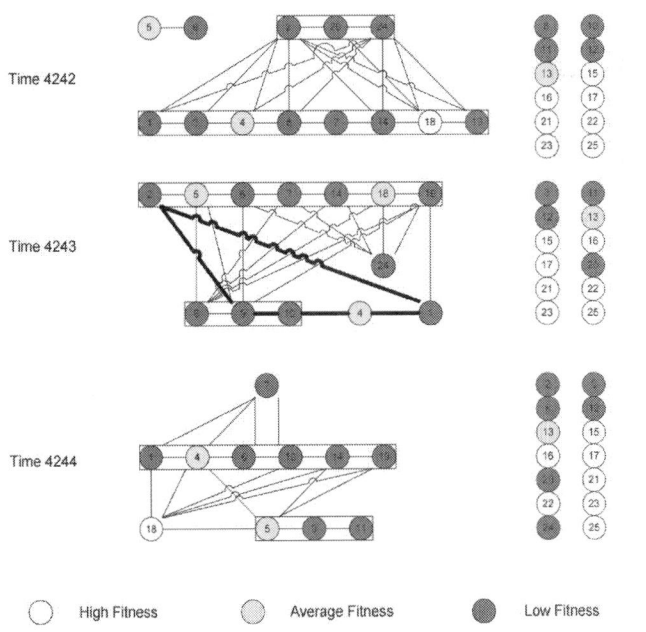

Figure 12: Dynamic q-clusters in the interaction poset.

Dynamic clusters regularly appear in simulation runs. We have found that the participants in these clusters typically have lower than average fitness values. In other words, they form "frozen cores" of agents within the agent set, around which the other agents are more or less free to move. Since this typically happens for relatively high P values, delay events have a relatively high impact, and so agents will tend to freeze into permanent states of delay, and the clusters expand by attracting additional agents.

However, we have observed that, with very high frequency, these clusters are prevented from becoming permanently frozen by the evolution of nontrivial elements of the A_1-groups of the associated covering complexes; these act to free up agents. Fig. 12 contains such an example. Agent 18 has high fitness at time step 4242, and A_1 of the covering complex of the poset representing the interaction at time step 4242 is trivial. Then, in time step 4243, Agent 18 has become part of a large cluster, which reduced its fitness. At the same time A_1 of the order complex at step 4243 has a nontrivial element, with a representative indicated with bold edges, involving Agents 2, 9, 10, 4, and 1. In time step 4244, Agent 18 has left the dynamic cluster again, has regained its high fitness, and the A1-group in time step 4244 is back to being trivial. Fig. 13 shows the parameter ranges that produce this phenomenon.

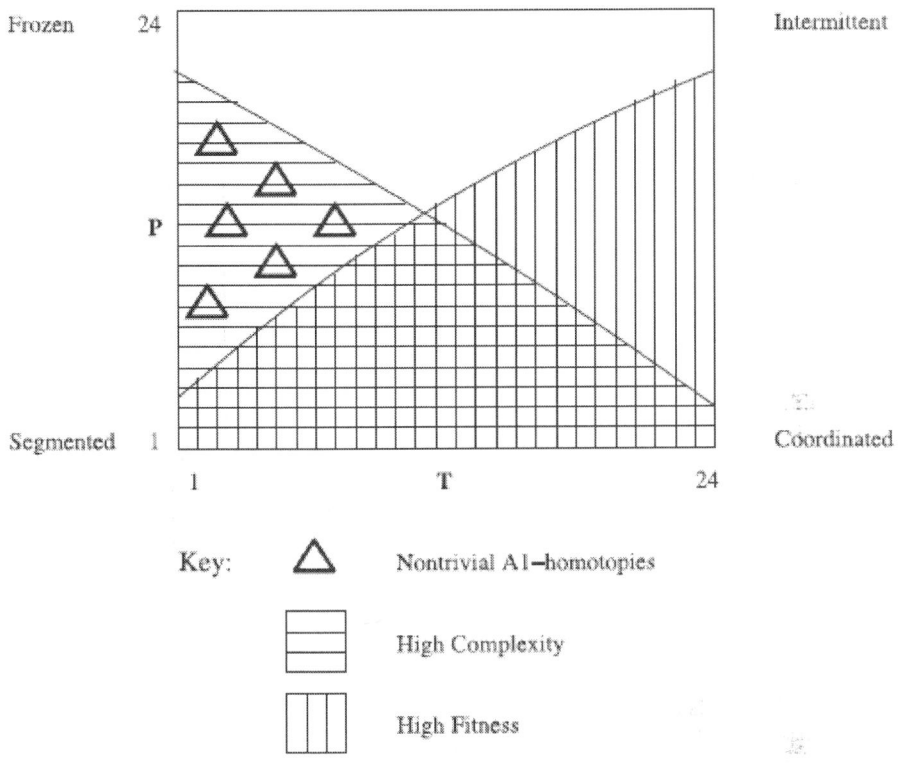

Figure 13: Parameter space.

At present we do not have an explanation for this phenomenon and are unable to state and prove precise results about it. We view this as an important open research question.

ACKNOWLEDGMENTS

The authors would like to thank the referees for their helpful comments and suggestions and for the patience of one of the referees with the first author's written English.

REFERENCES

1. R. Atkin An algebra for patterns on a complex, I Internat. J. Man–Machine Studies, 6 (1974), pp. 285–307
2. R. Atkin An algebra for patterns on a complex, II Internat. J. Man–Machine Studies, 8 (1976), pp. 448–483
3. R. Atkin Multidimensional Man Penguin, London (1981)
4. E. Babson, H. Barcelo, M. de Longueville, R. Laubenbacher, A Homotopy Theory for Graphs, preprint, 2003.
5. E. Babson, H. Barcelo, R. Laubenbacher, A-homotopy theory and arrangements of linear subspaces, in preparation.
6. H. Barcelo, X. Kramer, R. Laubenbacher, C. Weaver Foundations of a connectivity theory for simplicial complexes Adv. in Appl. Math., 26 (2001), pp. 97–128
7. A. Björner, Subspace arrangements, Proceedings of the First European Congress of Mathematics, vol. I, Paris, 1992, Progr. Math., ed., vol. 119, Birkhäuser, Basel, 1994, pp. 321–370.
8. A. Björner, Topological methods, in: Handbook Of Combinatorics, vol. II, MIT Press, North-Holland, Cambridge, MA, Amsterdam, 1995, pp. 1819–1872.
9. A. Björner, personal communication, 2001.
10. A. Björner, B. Korte, L. Lovász Homotopy properties of greedoids Adv. in Appl. Math., 6 (1985), pp. 447–494
11. A. Björner, L. Lovász Linear decision trees, subspaces arrangements, and Möbius functions J. Amer. Math. Soc., 7 (1994), pp. 677–706
12. A. Björner, L. Lovász, A. Yao, Linear decision trees: volume estimates and topological bounds, Proceedings, 24th ACM Symposium on Theory of Computing, New York, ACM, New York, 1992, pp. 170–177.
13. A. Björner, V. Welker The homology of "k-equal" manifolds and related partitions lattices Adv. in Math., 110 (1995), pp. 277–313
14. A. Björner, G. Ziegler Combinatorial stratification of complex arrangements J. Amer. Math. Soc., 5 (1) (1992), pp. 105–149
15. J.A. Bondy, Pancyclic graphs. II, in: R.C. Mullin, (Ed.), Proceedings of the Second Louisiana Conference on Combinatorics, Graph Theory and Computing, Baton Rouge, Louisiana, 1971 (summary).
16. C. De Concini, C. Procesi Wonderful models of subspaces arrangements Selecta Math., I (1995), pp. 459–494
17. R.L. Cummins Hamiltonian circuits in tree graphs IEEE Trans. Circuit Theory, 13 (1966), pp. 82–90
18. A. Frank, Some polynomial algorithms for certain graphs and hypergraphs, Proceedings of the Fifth British Combinatorial Conference, Aberdeen, 1975, Congr. Numerantium XV, 1976, pp. 221–226.
19. C.A. Holzmann, F. Harary On the tree graph of a matroid SIAM J. Appl. Math., 22 (1972), pp. 187–193

20. J. Johnson The mathematical revolution inspired by computing, The Mathematics of Complex Systems Oxford University Press, Oxford (1991)
21. M. Joswig Projectivities in simplicial complexes and colorings of simple polytopes Math. Z., 240 (2) (2002), pp. 243–259
22. X. Kramer, R. Laubenbacher, Combinatorial homotopy of simplicial complexes and complex information networks, in: D. Cox, B. Sturmfels (Eds.), Applications of Computational Algebraic Geometry, vol. 53, Proceedings of the Symposium in Applied Mathematics, American Mathematical Society, Providence, RI, 1998.
23. M. De Longueville, V. Reiner, personal communication, 2001.
24. L. Lovász A homology theory for spanning trees of a graph Acta Math. Acad. Sci. Hungar., 30 (3–4) (1977), pp. 241–251
25. G. Malle A homotopy theory for graphs Glasnik Mathematicki, 18 (38) (1983), pp. 3–25
26. S.B. Maurer, Problem presented, at the Fifth British Combinatorial Conference, Aberdeen, 1975.
27. S.B. Maurer Matroid basis graphs, I J. Combin. Theory Ser. B, 14 (1973), pp. 240–261
28. H. Shank A note on Hamiltonian circuits in tree graphs IEEE Trans. Circuit Theory, 15 (1968), p. 86
29. N.J.A. Sloane, On-line encyclopedia of integer sequences, http://www.research.att.com/njas/sequences/
30. W.T. Tutte A homotopy theorem for matroids, I–II Trans. Amer. Math. Soc., 88 (1958), pp. 144–174
31. G.M. Ziegler Lectures on polytopes, Graduate Texts in Mathematics Springer, Berlin (1994)

CITATION

Hélène Barcelo, Reinhard Laubenbacher, Perspectives on A-homotopy theory and its applications, Discrete Mathematics, Volume 298, Issues 1–3, 6 August 2005, Pages 39-61, ISSN 0012-365X, http://dx.doi.org/10.1016/j.disc.2004.03.016.

A Discrete Model of S^1-Homotopy Theory

Andrew J. Blumberg
Department of Mathematics, University of
Chicago, 5734 S. University Ave., Chicago, IL
60637, United States

ABSTRACT

We construct a discrete model of the homotopy theory of S^1-spaces. We define a category P with objects composed of a simplicial set and a cyclic set along with suitable compatibility data. P inherits a model structure from the model structures on the categories of simplicial sets and cyclic sets. We then show that there is a Quillen equivalence between P and the model category of S^1-spaces in which weak equivalences and fibrations are maps inducing weak equivalences and fibrations on passage to all fixed point sets.

INTRODUCTION

Simplicial techniques are often unavailable in the context of equivariant homotopy theory. When G is not a discrete group, simplicial G-sets do not provide a model for the homotopy theory of G-spaces. The lack of an adequate replacement for simplicial sets is a substantial inconvenience. Cyclic sets [3] provide a useful discrete model of a portion of S^1-homotopy theory. Specifically, Spalinski [15] (following Dwyer et al. [4]) constructs a model structure on cyclic sets which is Quillen equivalent to the model structure on S^1-spaces in which weak equivalences and fibrations are detected on passage to fixed point subspaces for finite groups. However, since the S^1-fixed points of the geometric realization of a cyclic set must be discrete [7] and [15], it is unreasonable to expect a model structure on cyclic sets which will capture all of S^1-homotopy theory.

Restating this observation, the category of cyclic sets encodes all of S^1-homotopy theory except for the information detected by the S^1-fixed points. A fundamental insight of Elmendorf [5] is that the homotopy theory of G-spaces is equivalent to the homotopy theory of appropriate diagrams of fixed point information. See also Mandell and Scull [12] for a comprehensive modern discussion of this. This suggests that a natural avenue of attack is to consider a category consisting of a cyclic set appropriately coupled (via compatibility data) with a simplicial set to represent the information at the S^1-fixed points. Let X be an S^1-space, and consider the following diagram:

$$X^{S^1} \times E\mathscr{F} \longrightarrow X \times E\mathscr{F}$$
$$\downarrow$$
$$X^{S^1}.$$

$$(1.1)$$

Here $E\mathscr{F}$ is the classifying space for the family of finite subgroups of S^1, the horizontal map is the inclusion and the vertical map is the projection. The associated pushout is weakly equivalent to X. This picture provides the inspiration for our construction. We think of the cyclic set as akin to $X \times E\mathscr{F}$, the simplicial set as X^{S^1}, and the compatibility data as the gluing along $X^{S^1} \times E\mathscr{F}$.

Given a simplicial set A and a cyclic set B we will describe the required compatibility in terms of a map $\nabla A \to B$, where ∇A is a homotopical cyclic approximation of A. Let $|-|_s$ and $|-|_c$ denote the geometric realization functors in the category of simplicial sets and cyclic sets respectively. We construct a functor $\nabla : S \to S^c$ which has the property that there is a natural map $|\nabla A|_c \to |A|_s$, which is a weak equivalence upon passage to all fixed point sets for finite subgroups of S^1. The category P of compatible pairs is an instance of a more general construction.

Definition 1.2
Let C and D be categories and $F: C \to D$ a functor. The category $_{CF}D$ has

1) Objects specified by triples $(A, B, FA \to B)$ where A is an object of C and B is an object of D.
2) Morphisms specified by maps $_{f_1}:A \to A'$ and $_{f_2}:B \to B'$ such that the following diagram commutes:

$$FA \longrightarrow B$$

$$Ff_1 \downarrow \qquad\qquad \downarrow f_2$$

$$FA' \longrightarrow B'.$$

$$(1.3)$$

Remark 1.4
This is an example of a comma category [10].

Definition 1.5

The category P is the comma category $S \nabla S^c$.
When there are model structures on C and D, there is an induced model structure on $_{CF}D$ for suitable functors F.

Definition 1.6
Let C and D be model categories. A functor $F:C \to D$ is Reedy admissible   if F preserves colimits (e.g. F is a left adjoint) and F has the property that given a morphism $(A, B, FA \to B) \to (A', B', FA' \to B')$ in $_{CF}D$ such that $A \to A'$ is a trivial cofibration in C and $FA'_{\cup FA} B \to B'$ is a trivial cofibration in D then $B \to B'$ is a weak equivalence in D (e.g. F is a left Quillen functor).

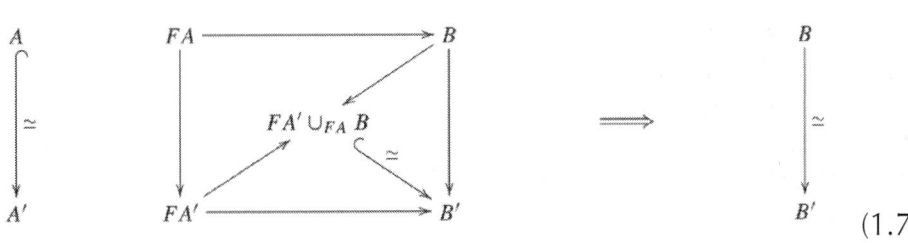

$$(1.7)$$

Theorem 1.1

Let C and D be model categories and $F:C \to D$ be a Reedy admissible functor. Then $_{CF}D$ admits a model structure. A map $(A, B, FA \to B) \to (A', B', FA' \to B')$ is

1) a weak equivalence if $A \to A'$ is a weak equivalence in C and $B \to B'$ is a weak equivalence in D,
2) a fibration if $A \to A'$ is a fibration in C and $B \to B'$ is a fibration in D,
3) a cofibration if $A \to A'$ is a cofibration in C and $FA'_{UFA}B \to B'$ is a cofibration in D.

We use this theorem to obtain the model structure on P.

Lemma 1.8

The functor $\nabla : S \to S^C$ is Reedy admissible. Type equation here.

Corollary 1.9

There is a model structure on P inherited from the model structures on S and S^c.

There is an adjunction specified by functors $L : \mathcal{P} \to \text{top}^{S^1}$ and $R : \text{Top}^{S^1} \to \mathcal{P}$. L is constructed from the realization functors $_{|-|S}$ and $_{|-|c'}$ and R is constructed from the singular functors S and $_{Sc}$.

Definition 1.10

The functor $L : \mathcal{P} \to \text{Top}^{S^1}$ takes a triple $(A, B, \nabla A \to B)$, to the pushout in the diagram:

$$
\begin{array}{ccc}
|\nabla A|_c & \longrightarrow & |B|_c \\
\downarrow & & \downarrow \\
|A|_s & \longrightarrow & X.
\end{array}
\tag{1.11}
$$

The functor $R : \text{Top}^{S^1} \to \mathcal{P}$ takes X to the triple $(S(X^{S1}), _{Sc}(X), \nabla S(X^{S1}) \to _{Sc}(X))$. The map $\xi : \nabla S(X^{S1}) \to _{Sc}(X)$ is the adjoint of the composite

$$|\nabla S(X^{S^1})|_c \to |S(X^{S^1})|_s \to X^{S^1} \hookrightarrow X. \tag{1.12}$$

Recall that there is a model structure on Top^{Si} given by defining a map $f: X \to Y$ to be a weak equivalence if $f^H: X^H \to Y^H$ is a weak equivalence for all $H \subset S^1$, a fibration if $f^H: X^H \to Y^H$ is a fibration for all $H \subset S^1$, and a cofibration if it has the left-lifting property with respect to acyclic fibrations [15].

Here is the main theorem of the paper.

Theorem 1.2
The functors L and R specify a Quillen equivalence between P with

the model structure given by Theorem 1.1 and Top^S with the model structure described above.

The problem of obtaining a discrete model for S^1-spaces was raised by Voevodsky in a 2002 e-mail to May [13].

The rest of the paper is organized as follows:

1) A brief review of simplicial and cyclic sets.
2) A review of the model structures on S^1 spaces.
3) Definition of ∇ and demonstration that it is Reedy admissible.
4) The adjunction between P and Top^S .
5) Proof of Theorem 1.2.
6) Proof of Theorem 1.1.

A REVIEW OF CYCLIC SETS

We give a very succinct review of cyclic sets. Good references for readers unfamiliar with the category are [3], [4], [7] and [15]. A cyclic set can be regarded as a simplicial set with extra data, namely an action of $Z/(n+1)$ on the n-simplices which is compatible with the face and degeneracy operators in the following sense. Define the

cyclic category Λ^{op} to have the same objects as the category Δ^{op} and the same generating morphisms along with an extra degeneracy

$s_{n+1}:[n]\to[n+1]$ and the "cyclic relations" $(d_0 s_{n+1})^{n+1}=\text{id}$. Define $t_n = d_0 s_{n+1}:[n]\to[n]$. Every morphism in Λ^{op} can be written as a composite $\emptyset = STD$ of a composite S of degeneracy operators, a power T of t_n for some n, and a composite D of boundary operators [15]. For further discussion of the properties of Λ^{op} (e.g. Λ^{op} is self-dual) see [3] or [6].

Cyclic sets are contravariant functors from the category Λ^{op} to sets. The category of cyclic sets will be denoted as S^c. As in the theory of simplicial sets, the represented cyclic sets $\Lambda[n]=\hom_\Lambda(-,n)$ play an important role. The geometric realization of the underlying simplicial set of a cyclic set admits a natural S^1-action. The geometric realization, regarded as a functor from cyclic sets to S^1-spaces, will be denoted as $|-|_c$. In particular, $|\Lambda[n]|_c\cong S^1\times|\Delta[n]|$, with S^1 acting on the product by rotation on the first coordinate, where $\Delta[n]$ is the represented simplicial set with trivial action. By manipulation of coends one obtains $|X|_c = X\otimes_{\Lambda op}|\Lambda|_c$. The adjoint to the realization is the "cyclic singular functor" S_c defined to have n-simplices $\hom_{S^1}(|\Lambda[n]|_c,X)$ [4]. Here the cyclic structure is obtained by regarding $|\Lambda[n]|_c$ as homeomorphic to $S^1\times|\Delta[n]|$, where the action of t_n permutes the coordinates of a point in $|\Delta[n]|$ and rotates S^1 by $e^{2ri/(n+1)}$.

Now consider the subgroup $Z/(r)\subset S^1$. Given a cyclic set, we can apply the subdivision functor Sd_r to the underlying simplicial set [1]. This has a natural simplicial action of $Z/(r)$ induced from the cyclic structure, and so we can define a composite functor Φ_r which takes a cyclic set X to the simplicial set $(Sd_r X)^{Z/(r)}$. There is a homeomorphism $|\Phi r(X)|_s\cong|X|^{Z/(r)}$. By Freyd's adjoint functor theorem, Φ_r has a left adjoint ψ_r. It is useful to describe ψ_r more concretely and so we reproduce calculations of Spalinski [14] inAppendix A.

The functors Φ_r are used to prove the following result.

Lemma 2.1 Spalinski [15]
The counit of the adjunction between $|-|_c$ and $S_c(-)$ induces weak equivalences on passage to fixed point spaces for finite subgroups of S^1.

A REVIEW OF MODEL STRUCTURES ON SIMPLICIAL SETS, CYCLIC SETS, AND TOP^{S1}

We briefly review the model structures on S, S^c, and Top^{S^1}.

Theorem 3.1
There is a model structure on simplicial sets in which a map is

1) a fibration if it is a Kan fibration,
2) a weak equivalence if the induced map on passage to geometric realization is an equivalence,
3) a cofibration if it is an injection.

Theorem 3.2

For any family $\mathcal{F}$ of subgroups of S^1, there is a model structure on Top^{S^1} in which a map $f : X \to Y$ is

1) a fibration if the induced maps $f^H : X^H \to Y^H$ are fibrations for all $H \in \mathcal{F}$,
2) a weak equivalence if the induced maps $f^H : X^H \to Y^H$ are weak equivalences for all $H \in \mathcal{F}$,
3) a cofibration if it has the left-lifting property with respect to trivial fibrations.

In particular, this holds when $\mathcal{F}$ is the family of all subgroups of S^1 and when $\mathcal{F}$ is the family of all finite subgroups of S^1.

Theorem 3.3 Spaliński [15]
There is a model structure on cyclic sets in which a map is

1) a fibration if $\Phi_r(f)$ is a fibration of simplicial sets for all $r \geq 1$,

2) a weak equivalence if $\Phi_r(f)$ is a weak equivalence of simplicial sets for all $r \geq 1$,
3) a cofibration if it has the left-lifting property with respect to trivial fibrations.

Remark 3.1
Spalinski [15] also shows that cofibrations can be characterized as retracts of transfinite composites of pushouts of coproducts of maps $\Psi_r(\partial\Delta[k]) \to \Psi_r(\Delta[k])$.
The homotopy theory of cyclic sets is the same as the homotopy theory

of Top^{S^1} with respect to the family of finite subgroups of S^1.

Theorem 3.4 Spalinski [15]
The cyclic realization functor and the cyclic singular functor induce a

Quillen equivalence between Top^{S^1} with the model structure in which $\mathcal{F}$ is the family of finite subgroups of S^1 and the category of cyclic sets with the model structure described above.

THE FUNCTOR ∇

We shall construct a functor $\nabla : S \to S^c$ such that $|X|_s^H \simeq |\nabla X|_c^H$ for all finite $H \subset S^1$. One's first guess is that ∇ ought to be the left adjoint to the forgetful functor which assigns to a cyclic set its underlying simplicial set (Kan extension). However, this is the free cyclic set associated with the underlying simplicial set [2], and does not have the properties we need.

Another obvious guess is to define $\nabla X = S_c(|X|_s)$. By Lemma 2.1, we know the counit provides a map $|S_c(|X|_s)|_c \to |X|_s$ which is an equivalence on passage to all finite subgroups. Unfortunately, as a composite of a left adjoint and a right adjoint, this functor has rather unpleasant properties. For instance, it preserves neither colimits nor limits.

We want a functor from simplicial sets to cyclic sets which is a left adjoint and so preserves colimits. All such functors arise from cosimplicial cyclic sets. In fact, there is an equivalence between the category of cosimplicial objects in C and adjunctions from simplicial sets to C for categories C with all small colimits [9, 3.1.5].

Definition 4.1

Set $\nabla_n = S_c(|\Delta[n]|)$. Then ∇_* is a cosimplicial cyclic set and so we can define a functor $\nabla : S \to S^c$ by letting $\nabla X = X \otimes \Delta^{op}\nabla_*$. The functor ∇ has the right adjoint $A : S^c \to S$ specified by $Z(Y)_n = \hom(\nabla_n, Y),..$ We will repeatedly use the following result, which we quote from [11].

Lemma 4.2

The functor $(-)^G$ on based G-spaces preserves pushouts of diagrams one leg of which is a closed inclusion.

Lemma 4.3

There is a natural map $\zeta : _{|\nabla X|_c} \to _{|X|_s}$ which induces weak equivalences on passage to fixed point subspaces for all finite subgroups of S^1.

Proof

By construction, the counit map $\gamma_n : |\nabla_n|_c \to _{|\Delta[n]|_s}$ induces weak equivalences on passage to all fixed point subspaces for finite subgroups of S^1. Define ζ to be the following map:

$$|\nabla X|_c = ((X \otimes_{\Delta^{op}} \nabla) \otimes_{\Delta^{op}} |\Delta|_s) = (X \otimes_{\Delta^{op}} (\nabla \otimes_{\Delta^{op}} |\Delta|_s)) = X \otimes_{\Delta^{op}} |\nabla|_c \longrightarrow X \otimes_{\Delta^{op}} |\Delta|_s = |X|_s.$$

Both the domain and the codomain can be regarded as a succession of pushouts with one leg a cofibration. Therefore the fixed point functor commutes with each of these coends by Lemma 4.2 and so ζ induces weak equivalences on passage to fixed subspaces.

Remark 4.4

There are two essential aspects of the ∇_n. First, they come equipped with maps from $|\nabla_n|_c$ to $|\Delta[n]|_s$ which induce weak equivalences on passage to fixed point subspaces for all finite subgroups of S^1. Second, the natural map $\mathrm{colim}_{i \to n} \nabla_i \to \nabla_n$ is an injection. Any other cosimplicial cyclic set which had these properties would suffice for our purposes. One might prefer a functorial cofibrant approximation of

∇_*. Alternatively, as the singular construction we give is rather bloated, we expect that other explicit models of ∇_n may well be preferable for specific applications.

To use Theorem 1.1 to show that there is a model structure on P, we must verify that ∇ is Reedy admissible. By construction, ∇ is a left adjoint and so preserves colimits.

Lemma 4.5

Given a map $(A, B, \nabla A \to B) \to (A', B', \nabla A' \to B')$ in P such that $A \to A'$ is a trivial cofibration and $\nabla A' \cup_{\nabla A} B \to B'$ is a trivial cofibration, the map $B \to B'$ is a weak equivalence. Therefore ∇ is Reedy admissible.

Proof

Since $B \to B'$ is the composite

$$B \to \nabla A' \cup_{\nabla A} B \to B' \tag{4.6}$$

and $\nabla A' \cup_{\nabla A} B \to B'$ is a weak equivalence by hypothesis, it suffices to show that $B \to \nabla A'_{\cup \nabla A} B$ is a weak equivalence. This is equivalent to showing that $|B|_c^H \to |\nabla A' \cup \nabla_A B|_c^H$ is a weak equivalence of spaces for all finite $H \subset S^1$. Geometric realization is a colimit, so $|\nabla A' \cup \nabla_A B|_c^H$ is isomorphic to $(|\nabla A'|_c \cup |\nabla A|_c |B|_c)^H$. Since $A \to A'$ is a cofibration of simplicial sets (and hence an inclusion), $|\nabla A|_c \to |\nabla A'|_c$ is a closed inclusion. Therefore by Lemma 4.2 the fixed point functor commutes with the pushout, and so $(|\nabla A'|_c \cup |\nabla A|_c |B|_c)^H$ is equivalent to $|\nabla A'|_c^H \cup_{|\nabla A|_c^H} |B|_c^H$. Finally $|\nabla A|_c^H \to |\nabla A'|_c^H$ is a trivial cofibration when $A \to A'$ is a trivial cofibration, so $|B|_c^H \to |\nabla A'|_c^H \cup_{|\nabla A|_c^H} |B|_c^H$ is the pushout of a trivial cofibration and is thus itself a trivial cofibration.

Corollary 4.7

There is a model structure on P in which a map is

(1) a weak equivalence if $A \to A'$ is a weak equivalence of simplicial sets and $B \to B'$ is a weak equivalence of cyclic sets,
(2) a fibration if $A \to A'$ is a fibration of simplicial sets and $B \to B'$ is a fibration of cyclic sets,

(3) a cofibration if $A \rightarrow A'$ is a cofibration of simplicial sets and $\nabla A' \cup_{\nabla_A} B \rightarrow B'$ is a cofibration of cyclic sets.

Proof
This follows immediately from Lemma 4.5 and Theorem 1.1. □

THE ADJUNCTION BETWEEN P AND TOP^{S1}

There are natural functors from P to Top^{S^1} and from Top^{S1} to P defined as follows.

Lemma 5.1
Given a morphism in P from $(A, B, \nabla A \rightarrow B)$ to $(A', B', \nabla A' \rightarrow B')$, the induced diagram

$$
\begin{array}{ccc}
|\nabla A|_c & \xrightarrow{\zeta} & |A|_s \\
\downarrow & & \downarrow \\
|\nabla A'|_c & \xrightarrow{\zeta} & |A'|_s
\end{array}
\qquad (5.2)
$$

is commutative.

Proof
Rewriting the diagram as follows

$$
\begin{array}{ccc}
A \otimes_{\Delta^{\text{op}}} |\nabla| & \longrightarrow & A \otimes_{\Delta^{\text{op}}} |\Delta| \\
\downarrow & & \downarrow \\
A' \otimes_{\Delta^{\text{op}}} |\nabla| & \longrightarrow & A' \otimes_{\Delta^{\text{op}}} |\Delta|
\end{array}
\qquad (5.3)
$$

makes the commutativity apparent.

Definition 5.4

The functor $L : P \to \mathrm{Top}^{S^1}$ takes a triple $(A, B, \nabla A \to B)$, to the pushout in the diagram:

$$
\begin{array}{ccc}
|\nabla A|_c & \longrightarrow & |B|_c \\
{\scriptstyle \zeta} \downarrow & & \downarrow \\
|A|_s & \longrightarrow & X.
\end{array}
\tag{5.5}
$$

A morphism $(A, B, \nabla A \to B) \to (A', B', \nabla A' \to B')$ induces a commutative diagram:

$$
\begin{array}{ccc}
|A|_s & \longleftarrow |\nabla A|_c \longrightarrow & |B|_c \\
\downarrow & \downarrow & \downarrow \\
|A'|_s & \longleftarrow |\nabla A'|_c \longrightarrow & |B'|_c.
\end{array}
\tag{5.6}
$$

The left-hand square commutes by the preceding lemma and the right-hand square commutes because of the definition of a morphism. Therefore there is an induced map of pushouts, which specifies the action of L on morphisms.

Lemma 5.7

A morphism $X \to Y$ in Top^{S^1} induces a commutative diagram:

$$
\begin{array}{ccc}
\nabla S(X^{S^1}) & \overset{\xi}{\longrightarrow} & S_c(X) \\
\downarrow & & \downarrow \\
\nabla S(Y^{S^1}) & \overset{\xi}{\longrightarrow} & S_c(Y).
\end{array}
\tag{5.8}
$$

Proof

This diagram commutes if and only if the adjoint diagram

$$|\nabla S(X^{S^1})|_c \longrightarrow X$$
$$\downarrow \qquad\qquad \downarrow$$
$$|\nabla S(Y^{S^1})|_c \longrightarrow Y \qquad\qquad (5.9)$$

commutes. The latter diagram can be written as the composite:

$$|\nabla S(X^{S^1})|_c \longrightarrow |S(X^{S^1})|_s \longrightarrow X^{S^1} \longrightarrow X$$
$$\downarrow \qquad\qquad \downarrow \qquad\qquad \downarrow \qquad\qquad \downarrow$$
$$|\nabla S(Y^{S^1})|_c \longrightarrow |S(Y^{S^1})|_s \longrightarrow Y^{S^1} \longrightarrow Y. \qquad (5.10)$$

Here the left-hand square commutes by Lemma 5.1, the middle square commutes by the naturality of the counit, and the right-hand square commutes trivially. Therefore the original diagram commutes.

Definition 5.11

The functor $R : \mathrm{Top}^{S^1} \to P$ takes X to the triple $(S(X^{S^1})), S_c(X)\nabla S(X^{S^1})$ $\to S_c(X))$. The map $\nabla S(X^{S^1}) \to S_c(X)$ is the adjoint of the composite:

$$|\nabla S(X^{S^1})|_c \to |S(X^{S^1})|_s \to X^{S^1} \hookrightarrow X. \qquad (5.12)$$

A map $X \to Y$ in Top^{S^1} induces maps $S(X^{S^1}) \to S(Y^{S^1})$ and $S_c(X) \to S_c(Y)$ by functoriality. By the preceding lemma, these maps fit into a commutative diagram:

$$\begin{array}{ccc}
\nabla S(X^{S^1}) & \longrightarrow & S_C(X) \\
\downarrow & & \downarrow \\
\nabla S(Y^{S^1}) & \longrightarrow & S_C(Y).
\end{array}$$

(5.13)

We think of L as a realization functor and R as a singular functor.

Proposition 5.14
The functors

$$L : \mathscr{P} \rightleftarrows \mathrm{Top}^{S^1} : R$$

(5.15)

form an adjoint pair.

Proof
Given a map $|A|_s \cup |\nabla A|_c |B|_c \to X$, we must show that there is a unique corresponding map$(A, B, \nabla A \to B) \to (S(X^S), S_c(X), \nabla S(X^S) \to S_c(X))$. We clearly get unique maps $A \to S(X)$ and $B \to S_c(X)$ as adjoints to the maps $|A|_s \to X$ and $|B|_c \to X$ induced by the map from the pushout. It suffices to verify that the compatibility imposed by the pushout square is equivalent to the compatibility condition for a morphism in P.
So consider the square induced by our adjoint maps:

$$\begin{array}{ccc}
\nabla A & \longrightarrow & B \\
\downarrow & & \downarrow \\
\nabla S(X^{S^1}) & \longrightarrow & S_C(X).
\end{array}$$

(*)

We must show that it commutes. Now, the map $|A|_s \cup |\nabla A|_c |B|_c \to X$ provides us with a commuting square:

$$|\nabla A|_c \longrightarrow |B|_c$$
$$\downarrow \qquad\qquad \downarrow$$
$$|A|_s \longrightarrow X. \tag{5.16}$$

Such squares are in bijective correspondence with commuting squares:

$$\nabla A \longrightarrow B$$
$$\downarrow \qquad\qquad \downarrow$$
$$S_c(|A|_s) \longrightarrow S_c(X). \tag{$**$}$$

The two composites $\nabla A \to B \to S_c(X)$ are the same. Therefore to verify the correspondence of the compatibility conditions it suffices to show that the maps

$$(*) \quad \nabla A \to \nabla S(X^{S^1}) \to S_c(X) \qquad (**) \quad \nabla A \to S_c(|A|_s) \to S_c(X) \tag{5.17}$$

are identical. We do this by explicitly chasing elements around these two paths. Start with the map $g : |A|_s \to X$. Regarding $|A|_s$ as the coend $A \otimes_{\Delta^{op}} |\Delta|$, we view g as taking (a,Δ) to $g(a,\Delta)$ and its adjoint as taking a to the map $(\Delta \to g(a,\Delta))$.
So let's unwind the two maps. The map

$$\nabla A \to S_c(|A|_s) \to S_c(X) \tag{$**$}$$

Is the composition of

$$\nabla A \to S_c(|A|_s) \quad \text{and} \quad S_c(|A|_s) \to S_c(X). \tag{5.18}$$

The first constituent map is adjoint to the map $_{|\nabla A|_c} \to {}_{|A|_s}$ which we defined as

$$A \otimes_{\Delta^{op}} |\nabla|_c \to A \otimes_{\Delta^{op}} |\Delta|_s \tag{5.19}$$

via the map $\gamma:|\nabla|_c\to|\Delta|_s$. In order to calculate the adjoint map, we write the first coend as

$$(A \otimes_{\Delta^{\mathrm{op}}} \nabla) \otimes_{\Lambda^{\mathrm{op}}} |\Lambda|_c \to A \otimes_{\Delta^{\mathrm{op}}} |\Delta|_s$$

$$(5.20)$$

where the map takes $((a,v),\lambda)$ to $(a,\gamma(v,\lambda))$. Then the adjoint is the map

$$\lambda \to ((a, v) \to (a, \gamma(v, \lambda))).$$

$$(5.21)$$

Next, we have the map $S_c(|A|_s)\to_{S_c}(X)$ which is obtained by applying S_c to the map $g:|A|_s\to X$. That is, the induced map takes the map $\lambda\to(a,\Delta)$ to the map $\lambda\to g(a,\Delta)$. Finally, the composite is

$$(a, v) \to (\lambda \to g(a, \gamma(v, \lambda))).$$

$$(**)$$

On the other hand, we can decompose the map

$$\nabla A \to \nabla S(X^{S^1}) \to S_c(X)$$

$$(*)$$

as the composition of

$$\nabla A \to \nabla S(X^{S^1}) \quad \text{and} \quad \nabla S(X^{S^1}) \to S_c(X).$$

$$(5.22)$$

The first constituent map is obtained by applying ∇ to the map $A\to S(X^{S^1})$ adjoint to g. Explicitly, this is

$$(a, v) \to ((a \to (\delta \to g(a, \delta))), v).$$

$$(5.23)$$

The second map is the adjoint to the map $|\nabla S(X^{S^1})|_c\to X$, which decomposes as the composite

$$|\nabla S(X^{S^1})|_c \to |S(X^{S^1})|_s \to X^{S^1} \to X$$

$$(5.24)$$

that takes (h, v, λ) to $(h, \gamma(v, \lambda))$ and then to $h(\gamma(v, \lambda))$. The adjoint can be written as:

$$(h, v) \to (\lambda \to h(\gamma(v, \lambda))). \tag{5.25}$$

Composing, we have

$$(a, v) \to (\lambda \to g(a, \gamma(v, \lambda))). \tag{*}$$

PROOF OF THEOREM 1.2

The functors L and R are compatible with our model structures.

Lemma 6.1

Let P have the model structure described in Corollary 4.7 and Top^{S^1} have the model structure generated by the family of all subgroups of S^1. Then the adjoint functors L and R form a Quillen adjunction.

Proof
It suffices to show that R preserves fibrations and trivial fibrations. If $X \to Y$ is a fibration or a trivial fibration, then $S(X) \to S(Y)$ and $S_c(X) \to S_c(Y)$ are as well since $S_c(-)$ and $S(-)$ are themselves right Quillen adjoints.

Remark 6.2
In fact, R preserves weak equivalences since both $S(-)$ and $S_c(-)$ preserve weak equivalences.

Now, one potential problem with this model for Top^{S^1} is that while cyclic sets don't capture "useful" data at the S^1-fixed points, they do have some information there which might corrupt the data encoded in the simplicial set. In fact, it isn't in general the case that the counit $|S(X^{S^1})|_s \cup_{|\nabla S(X^{S^1})|_c} |S_c(X)|_c \to X$ is a weak equivalence of S^1-spaces. However, the following lemmas show that this map is an equivalence once we pass to cofibrant approximations. Observe that $(A, B, \nabla A \to B)$ cofibrant implies that A is cofibrant and $\nabla A \to B$ is a cofibration.

Lemma 6.3

If $X \to Y$ is cofibration of cyclic sets, then the induced map $\left|X\right|_c^{S^1} \to \left|Y\right|_c^{S^1}$ is a homeomorphism.

Proof

As noted previously, a cofibration of cyclic sets is a retract of a relative cell complex with respect to the family $\Psi_r(\partial\Delta[k]) \to \Psi_r(\Delta[k])$. A retract of a homeomorphism is a homeomorphism. Thus, it will suffice to observe that the domains and codomains of these generating cofibrations have no S^1-fixed points, as the fixed point functor commutes with these pushouts after passage to cyclic realization by Lemma 4.2. But this is true by an explicit calculation [14] which we reproduce in Appendix A.

Corollary 6.4

If X is a cofibrant cyclic set, then $\left|X\right|_c^{S^1}$ is empty.

Lemma 6.3 enables us to show that for cofibrant objects in P the gluing behaves properly.

Lemma 6.5

Let $(A, B, \nabla A \to B)$ be a cofibrant object in P and define $Z = \left|A\right|_s \cup_{\left|\nabla A\right|_c} \left|B\right|_c$. Then $Z^{S^1} \simeq \left|A\right|_s$ and for all finite $H \subset S^1$, $Z^H \simeq \left|B\right|_c^H$.

Proof

Observe that by Lemma 4.2, passage to fixed points commutes with the pushout since $\nabla A \to B$ is a cofibration. First consider the S^1-fixed points. We have a map $\left|A\right|_s \to Z^{S^1}$ induced by the pushout. Since $\nabla A \to B$ is a cofibration, Lemma 6.3 tells us that $\left|\nabla A\right|_c^{S^1} \simeq \left|B\right|_c^{S^1}$. But this immediately implies that $\left(_{\left|A\right|_s \cup \left|\nabla A\right|_c \left|B\right|_c}\right)^{S^1} \cong {}_{\left|A\right|_s}$. Now consider a finite subgroup $H \subset S^1$. Since $\left|\nabla A\right|_c^H \simeq \left|A\right|_s^H$ and $\left|\nabla A\right|_c^H \to \left|B\right|_c^H$ is a cofibration, $\left|B\right|_c^H \to Z^H$ is the pushout along a cofibration of a weak equivalence. Therefore $\left|B\right|_c^H \to Z^H$ is a weak equivalence since Top is proper.

Theorem 1.2

The functors L and R specify a Quillen equivalence between P with the model structure given by Theorem 1.1 and Top^{S^1} with the model structure in which $\mathcal{F}$ is the family of all subgroups of S^1.

Proof

We must show that given a cofibrant object $(A, B, \nabla A \to B)$ in P and a fibrant S^1-space X, a map $(A, B, \nabla A \to B) \to RX$ is a weak equivalence if and only if the adjoint $L(A, B, \nabla A \to B) \to X$ is a weak equivalence. Writing out the functors, we need to show that

$$(A, B, \nabla A \to B) \to (S(X^{S^1}), S_c(X), \nabla S(X^{S^1}) \to S_c(X)) \qquad (6.6)$$

is a weak equivalence if and only if

$$|A|_s \cup_{|\nabla A|_c} |B|_c \to X \qquad (6.7)$$

is a weak equivalence.

So assume that $|A|_s \cup_{|\nabla A|_c} |B|_c \to X$ is a weak equivalence. This implies that the induced map

$$(|A|_s \cup_{|\nabla A|_c} |B|_c)^{S^1} \to X^{S^1} \qquad (6.8)$$

is a weak equivalence. Furthermore, by Lemma 6.5 the map

$$|A|_s \to (|A|_s \cup_{|\nabla A|_c} |B|_c)^{S^1} \qquad (6.9)$$

is a weak equivalence, and so the composition is a weak equivalence. This implies that the adjoint $A \to S(X^{S^1})$ is a weak equivalence of simplicial sets. Similarly, for any finite $H \subset S^1$ the assumption implies that the induced map

$$(|A|_s \cup_{|\nabla A|_c} |B|_c)^H \to X^H \qquad (6.10)$$

is a weak equivalence and Lemma 6.5 tells us that

$$|B|_c^H \to (|A|_s \cup_{|\nabla A|_c} |B|_c)^H \qquad (6.11)$$

is a weak equivalence. Therefore the composite is a weak equivalence, and this implies that the adjoint $B \to S_c(X)$ is a weak equivalence of cyclic sets.

Conversely, assume that the adjoint

$$(A, B, \nabla A \to B) \to (S(X^{S^1}), S_c(X), \nabla S(X^{S^1}) \to S_c(X)) \tag{6.21}$$

is a weak equivalence. This implies that $|A|_s \to X^{S^1}$ is a weak equivalence of simplicial sets and that $|B|_c \to X$ is a weak equivalence of cyclic sets. The previous discussion and the "two out of three" property for weak equivalences now imply that $|A|_s \cup_{|\nabla A|c} |B|_c \to X$ is a weak equivalence.

PROOF OF THEOREM 1.1

The proof that $_{CF}D$ inherits a model structure from model structures on C and D when F is Reedy admissible uses the standard technique for lifting model structures to diagram categories indexed by Reedy categories [8] and [9].

Theorem 1.1
Let C and D be model categories and $F{:}C \to D$ be a Reedy admissible functor. Then $_{CF}D$ admits a model structure. A map $(A, B, FA \to B) \to (A', B', FA' \to B')$ is

1) a weak equivalence if $A \to A'$ is a weak equivalence in C and $B \to B'$ is a weak equivalence in D,
2) a fibration if $A \to A'$ is a fibration in C and $B \to B'$ is a fibration in D,
3) a cofibration if $A \to A'$ is a cofibration in C and $FA' \cup_{UFA} B \to B'$ is a cofibration in D.

Proof
1) $_{CF}D$ has all small limits and colimits since F preserves colimits and C and D have all small limits and colimits.
2) Weak equivalences satisfy the "two out of three" axiom since they do in C and D.

3) It is clear that the weak equivalences and fibrations are closed under retracts, since they are defined levelwise. We need to verify that retracts of cofibrations are cofibrations. The commutative diagram:

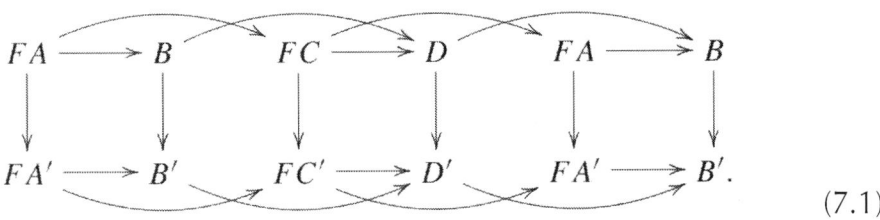

$$(7.1)$$

implies that $FC' \cup_{FC} D \to D'$ is a retract of $FA' \cup_{FA} B \to B'$. Since $FA' \cup_{FA} B \to B'$ is a cofibration, we know from the model structure on D that $FC' \cup_{FC} D \to D'$ is itself a cofibration in D. Moreover, it is clear that $C \to C'$ is a cofibration in C because it is a retract of $A \to A'$.

4) Now we need to verify the factorization results. Assume we have a map $(A, B, FA \to B) \to (A', B', FA' \to B')$. We will construct a factorization of this map into a trivial cofibration and a fibration (the other case is analogous). Consider the following diagram:

$$(7.2)$$

We employ the standard latching space argument. Choose a factorization of $A \to A'$ as $A \to C \to A'$ where $A \to C$ is a trivial cofibration in C and $C \to A'$ is a fibration. This yields a factorization $FA \to FC \to FA'$. So now we have the following diagram:

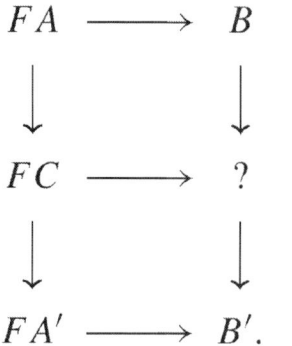

$$(7.3)$$

To complete the diagram choose a factorization of $FC \cup_{FA} B \to B'$ as

$$FC \cup_{FA} B \to C' \to B' \qquad (7.4)$$

where $C \cup_A B \to C'$ is a trivial cofibration and $C' \to B'$ is a fibration, and then put C' in for the "?". By the assumption on F, $B \to C'$ is a weak equivalence. This yields the factorization

$$(A, B, FA \to B) \to (C, C', FC \to C') \to (A', B', FA' \to B') \qquad (7.5)$$

in which the first arrow is a trivial cofibration and the second a fibration.

5) Finally, we must verify the lifting properties. Assume we have a trivial cofibration and a fibration (the other case is analogous). The lifting problem

$$(A, A', FA \to A') \longrightarrow (B, B', FB \to B')$$
$$\downarrow \qquad\qquad\qquad \downarrow$$
$$(X, X', FX \to X') \longrightarrow (Y, Y', FY \to Y') \qquad (7.6)$$

splits into the following interlocked lifting problems:

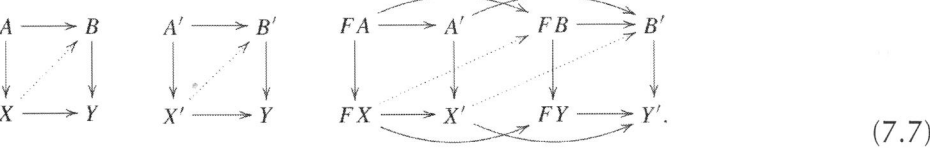

$$(7.7)$$

First, take a lift $X \to B$ in the left-hand diagram using the model structure on C. Now consider the diagram:

$$
\begin{array}{ccc}
FX \cup_{FA} A' & \longrightarrow & B' \\
\downarrow & & \downarrow \\
X' & \longrightarrow & Y'.
\end{array}
$$

$$(7.8)$$

Here the map $FX \cup_{FA} A' \to B'$ is built using the map $FX \to FB$ obtained from the lift. Take a lift $X' \to B'$ in this diagram using the model structure in D. Together, these two lifts provide the desired lifting.

Remark 7.9
There is a dual version of this result for categories with objects $(A, B, A \to GB)$ in which G is a co-Reedy admissible functor. That is, G preserves limits and satisfies an appropriate pullback condition.

Lemma 7.10
If C and D are left proper and F is Reedy admissible, then $_{CF}D$ is left proper. If C and D are right proper and F is a left Quillen functor, then $_{CF}D$ is right proper.

Proof
The first assertion follows since fibrations, weak equivalences, and pullbacks are defined levelwise. For the second assertion, we need that $(A, B, FA \to B) \to (A', B', FA' \to B')$ a cofibration implies that $B \to B'$ is a cofibration. If F is a left Quillen functor, this follows from [8, 15.3.11].

ACKNOWLEDGMENTS

I wish to express my deep gratitude to Mike Mandell for his generous assistance in solving this problem and to Peter May for his many useful comments and suggestions. I would also like to thank Phil Hirschhorn and Mark Hovey for helpful discussion and Jan Spalinski for permitting me to reproduce a portion of his thesis in Appendix A.

APPENDIX A. CALCULATIONS FROM SPALINSKI'S THESIS

We reproduce several calculations which appeared in Spalinski's thesis [14] but not in the paper based on the thesis [15].

A.1. Explicit Calculation of Ψ_r

We need to calculate $\Psi_r(\Delta[k])$. Recall that Ψ_r is the adjoint to Φ_r, where $\Phi_r(X) = (Sd_rX)^{\mathbb{Z}/(r)}$. It is sufficient to find a cyclic set A such that there is a natural equivalence

$$\hom_{S^c}(A, X) \rightarrow \hom_S(\Delta[k], \Phi_r(X)). \tag{A.1}$$

We know that there is an equivalence

$$\hom_S(\Delta[k], \Phi_r(X)) \rightarrow \Phi_r(X)_k \tag{A.2}$$

given by $f \mapsto f(\iota_k)$. So it will suffice to exhibit a cyclic set A such that there is a natural equivalence

$$\hom_{S^c}(A, X) \rightarrow \Phi_r(X)_k. \tag{A.3}$$

There is an action of $\mathbb{Z}/(n+1)$ on $|\Lambda[n]|_c$ for $n \geq 1$. By the Yoneda lemma, each map $\Lambda[n] \rightarrow \Lambda[n]$ is of the form $\hom_{\Lambda^{op}}(\phi, -1)$ for some $\phi : [n] \rightarrow [n] \in \Lambda^{op}$. The map corresponding to t_{n+1} has order $n+1$. This provides the action of $Z/(n+1)$, and we refer to the generator of this action as α. This action induces an action of $Z/(n+1)$ on $|\Lambda[n]|_c$.

Definition A.4
If k divides $n+1$, let $\Lambda[n|k]$ denote the orbit space of $\Lambda[n]$ with respect to the action of the subgroup of $Z/(n+1)$ generated by α^k.

Proposition A.5
The map

$$\hom_{S^c}(\Lambda[r(k+1)-1 \mid k+1], X) \to \Phi_r(X)_k \qquad (A.6)$$

given by $f \mapsto f[\iota_{r(k+1)}-1]$ is a bijection and so $\Psi_r(\Delta[n]) = \Lambda[r(n+1)-1|n+1]$.

Proof
First note that the image of γ is actually contained in the above fixed point set:

$$t_{r(k+1)}^{k+1} \cdot \gamma(f) = t_{r(k+1)}^{k+1} \cdot f([\iota_{r(k+1)-1}])$$
$$= f(t_{r(k+1)}^{k+1}[\iota_{r(k+1)-1}])$$
$$= f([\iota_{r(k+1)-1}]) = \gamma(f).$$

Next, observe that γ is onto. Take $x \in X_{r(K+1)-1}^H$ and consider the map:

$$f : \Lambda[r(k+1)-1] \to X, \qquad \iota_{r(k+1)-1} \mapsto x. \qquad (A.7)$$

Note that $\text{Im} f \subseteq X_{r(K+1)-1}^H$. Let $z = STD[t_{r(k+1)-1}] \in A[r(k+1)-1]$. We need to show that $f(\alpha_k + 1^z) = f(z)$. We have:

$$f(\alpha_{k+1}z) = f(\alpha_{k+1}STD[\iota_{r(k+1)-1}]) = f(STDt_{r(k+1)}^{k+1}[\iota_{r(k+1)-1}])$$
$$= STDt_{r(k+1)}^{k+1}f[\iota_{r(k+1)-1}] = STDt_{r(k+1)}^{k+1}x = STDx$$
$$= STDf(\iota_{r(k+1)-1}) = f(STD[\iota_{r(k+1)-1}]) = f(z).$$

Hence f factors as:

$$\Lambda[r(k+1)-1] \to \Lambda[r(k+1)-1]/\alpha^{k+1} \to X.$$

$$(A.8)$$

Here the first map is the quotient map and the second map is $\overline{f}$. By construction $\gamma(\overline{f}) = x$. Finally, we need to check that γ is injective. Suppose that $\gamma(f)=\gamma(g)$. Then $f[\iota_{r(k+1)-1}]=g[\iota_{r(k+1)-1}]$. Since $[\iota_{r(k+1)-1}]$ generates $\Lambda[r(k+1)-1]/\alpha^{k+1}$, $f=g$.

A.2. Fixed Points of $\Psi_r(\Delta[N])$

The explicit description of $\Psi_r(\Delta[n])$ makes it easy to calculate its S^1-fixed points.

Proposition A.9

For $k|(n+1)$, $|\wedge[n|k]|_c^{S^1} = \phi$.

Proof

Let $p\in(\Delta[n]\times S^1)$. Then we have

$$p = (x_0, x_1, \ldots, x_n, t) \quad x_i \geq 0, \sum_{i=0}^{n} x_i = 1, t \in S^1.$$

$$(A.10)$$

Let $\sigma=(0,1,\ldots,n)$, $\tau=\sigma^{-k}$, and $\gamma = e^{2\pi i/\frac{n+1}{k}}$. The action of α^k on $(\Delta[n]\times S^1)$ is given by

$$\alpha^k(x_0, x_1, \ldots, x_n, t) = (x_{\tau(0)}, x_{\tau(1)}, \ldots, x_{\tau(n)}, \gamma t).$$

$$(A.11)$$

Since S^1 is infinite and each orbit of α^k has only finitely many points,

$$\{(\Delta[n] \times S^1)/\alpha^k\}^{S^1} = \emptyset.$$

$$(A.12)$$

REFERENCES

1. M. Bokstedt, W.C. Hsiang, I. Madsen, The cyclotomic trace and algebraic K-theory of spaces, Invent. Math. 111 (1993) 465–539.
2. D. Burghelea, Z. Fiedorowicz, Cyclic homology and algebraic K-theory of spaces — II, Topology 25 (3) (1986) 303–317.
3. A. Connes, Cyclic homology and functors Extn , C. R. Acad. Sci. Paris 296 (1983) 953–958.
4. W.G. Dwyer, M.J. Hopkins, D.M. Kan, The homotopy theory of cyclic sets, Trans. Amer. Math. Soc. 291 (1985) 281–289.
5. A. Elmendorf, Systems of fixed point sets, Trans. Amer. Math. Soc. 277 (1983) 275–284.
6. A. Elmendorf, A simple formula for cyclic duality, Proc. Amer. Math. Soc. 118 (3) (1993) 709–711.
7. Z. Fiedorowicz, W. Gadja, The S 1 -CW decomposition of the geometric realization of a cyclic set, Fund. Math. 145 (1994) 91–100.
8. P. Hirschhorn, Model Categories and Their Localizations, in: Mathematical Surveys and Monographs, vol. 99, Amer. Math. Soc., Providence, RI, 2003.
9. M. Hovey, Model Categories, in: Mathematical Surveys and Monographs, vol. 63, Amer. Math. Soc., Providence, RI, 1999.
10. S. Mac Lane, Categories for the Working Mathematician, 2nd ed., in: Graduate Texts in Mathematics, vol. 5, Springer-Verlag, New York, NY, 1998.
11. M. Mandell, J.P. May, Equivariant Orthogonal Spectra and S-modules, Mem. Amer. Math. Soc. 159 (755) (2002).
12. M. Mandell, L. Scull, Algebraic models for equivariant homotopy theory over compact abelian lie groups, Math. Z. 240 (2) (2002) 261–287.
13. J.P. May, personal communication, 2003.
14. J. Spalinski, Strong homotopy theory of cyclic sets, Ph.D. Thesis, University of Notre Dame, 1993.
15. J. Spalinski, Strong homotopy theory of cyclic sets, J. Pure Appl. Algebra 99 (1995) 35–52.

CITATION

Andrew J. Blumberg, A discrete model of -homotopy theory, Journal of Pure and Applied Algebra, Volume 210, Issue 1, July 2007, Pages 29-41, ISSN 0022-4049, http://dx.doi.org/10.1016/j.jpaa.2006.08.011.

Derived Completions in Stable Homotopy theory

Gunnar Carlsson
Department of Mathematics, Stanford University,
Stanford, CA 94305, United States

5

ABSTRACT

We introduce a notion of derived completion applicable to arbitrary homomorphisms of commutative S-algebras, and work out some of its properties, including invariance results, a spectral sequence proceeding from purely algebraic information to the geometric results, and analysis of relationships with earlier constructions. We also provide some examples. The construction has applications in algebraic K-theory.

INTRODUCTION

It has long been recognized that the development of a theory of ring and module spectra, which bears the same relationship to the category of spectra as the ordinary theory of rings and modules has to the category of abelian groups, is a very desirable thing. A number of such theories exist. The different approaches [10] and [12] solve the problem in a satisfactory way, and more recently (see [14]) versions using orthogonal spectra and Γ-spaces are also available. The goal of transporting constructions which are available for ordinary rings and modules to this new category of ring spectra is also a very worthwhile one.

Some of these constructions have already been made by the authors of [10] and [12]. In this paper, we will use the notion of an S-algebra (as in [10]) as the spectrum version of a ring. Given an S-algebra A and module spectra M and N, one can construct spectra $M \wedge_A N$ and $\text{Hom}_A(M,N)$, analogous to the constructions $M \otimes_A N$ and $\text{Hom}_A(M,N)$ for rings A and modules M and N. From the point of view of topologists, the most important constructions to transport to the category of spectra are those which are homotopy invariant, i.e. those for which module or ring homomorphisms which induce weak equivalences on underlying spectra induce equivalences on the constructions. For this reason, one considers only derived constructions, i.e. constructions which on the algebraic side would always replace a module by a projective resolution for it, and would replace a ring by a level wise free simplicial ring. In the context of these categories of modules, this means that one replaces rings and modules by cofibrant and/or fibrant objects in the categories of S-algebras and module spectra. Our goal in this paper is to introduce and study the derived version of the completion construction for modules over a commutative ring.

We describe the main results. Let $f: A \rightarrow B$ be a homomorphism of commutative S-algebras, and let M be an A-module spectrum. Then we define a cosimplicial A-module spectrum $T_A(M;B)$, and the derived completion of M along the homomorphism f, denoted by $M_B^\wedge$, as the total spectrum of $T_A(M;B)$. Here are the properties of this construction that we will prove in this paper.

- The construction $M \rightarrow M_B^\wedge$ is functorial for homomorphism of module spectra over A.
- The construction is functorial in B in the sense that if $B \rightarrow C$ is a homomorphism of commutative A-algebras, we obtain an induced homomorphism $M_B^\wedge \rightarrow M_C^\wedge$.
- The construction is functorial in A in the following sense. Let $A \rightarrow B \rightarrow C$ be a diagram of commutative S-algebras, and let M be a B-module. Let $_{\rho A}(M)$ denote the spectrum regarded as an A-module spectrum by restriction of scalars along the ring homomorphism $A \rightarrow B$. Then there is an induced map $\rho_A M_C^\wedge \rightarrow M_C^\wedge)$.

- For any commutative ring A, A may be regarded as a commutative S-algebra via the Eilenberg–MacLane spectrum $\mathbb{H}(A)$, and an A-module M can be regarded as a module spectrum $\mathbb{H}(M)$. Thus, for a homomorphism of commutative rings $f: A \to B$, it is possible to construct the derived completion $\mathbb{H}(M)^\wedge_{\mathbb{H}(B)}$. For general A, B and M, it is possible that although $\mathbb{H}(A)$, $\mathbb{H}(B)$, and $\mathbb{H}(M)$ have no higher homotopy, the derived completion will. However, if A is Noetherian, M is finitely generated, and f is surjective, the derived completion of $\mathbb{H}(M)$ coincides with the Eilenberg–MacLane construction $\dfrac{\mathbb{H}(M^\wedge_I)}{\pi_* M^\wedge_B}$, where $M^\wedge_I$ is the usual algebraic completion construction at the ideal $I = \ker(f)$.

- One can show that in a sense, the completion "depends only on $\pi_0(B)$ for (-1)-connected S-algebras". The precise statement, which is our Theorem 6.1, is that if we are given a diagram $f: A \to B \to C$ of (-1)-connected commutative S-algebras with $\pi_0 B \to \pi_0 C$ an isomorphism, and an A-module spectrum M, then the map $M^\wedge_B \to M^\wedge_C$ is a weak equivalence of spectra.

- For any homomorphism $f: A \to B$ of (-1)-connected commutative S-algebras, and any connective A-module spectrum M (i.e. $\pi_s M = 0$ for s sufficiently small), there is a spectral sequence whose E_2-term depends only on the structure of $\pi_* M$ as a module over $\pi_0 A$ and of $\pi_0 B$ as an algebra over $\pi_0 A$, and which converges to $\pi_* M^\wedge_B$. This is our Theorem 7.1.

By far the most important use of the notion of completion in homotopy has been in p-adically or profinitely completing spaces. These constructions are in a sense "tame", in that for spectra with finitely generated homotopy groups, the homotopy groups of the completion can be obtained by algebraically completing the homotopy groups at primes or profinitely, and in general are viewed as a simplification of the homotopy type. When the homotopy groups are not finitely generated, one has situations where there is a single derived functor of completion which contributes to the homotopy groups of the completion. However, when applied to more complicated rings, our construction can construct interesting homotopy types from discrete rings, even for finitely generated modules over the ring. For example, let A be the group ring of the discrete group $\mathbb{Z}/p^{\infty}\mathbb{Z} = \bigcup_n \mathbb{Z}/p^n\mathbb{Z}$, and F_p as an A-algebra via augmentation followed by reduction mod p. If we form the derived completion of A itself as an A-module (using the Eilenberg–MacLane construction as above), we obtain the p-adically completed group ring on the singular chains on the circle group, regarded as a simplicial ring. What has in effect happened is that the derived completion construction on a discrete ring has created an S-algebra (which is much like a topological ring), which coincides with our geometric notion of "filling in the gaps" in the group $\mathbb{Z}/p^{\infty}\mathbb{Z}$, viewed as embedded in the circle group. So in this case, the completion produces interesting homotopy types related to embeddings of the ring within topological rings. This phenomenon has some similarity with the behavior of Quillen's plus construction, which replaces the classifying space of a discrete group with a homologically equivalent space. In [13], Lawson has studied the completion process in the context of the pro-p completion of a finitely generated nilpotent group Γ, and has shown that the homotopy groups of the completion are strongly related with the homology groups of stable (with respect to dimension) representation varieties for Γ.

We are developing this material for use in applications in algebraic K-theory, specifically in order to understand the descent problem for the K-theory of fields (see [6]). The goal is to obtain a homotopy theoretic model for the K-theory spectrum of a field F which depends only on the absolute Galois group G_F of the F. It turns out that a model which is often correct can be constructed out of the S-algebra associated with the symmetric monoidal category of finite dimensional complex representations of G_F precisely by performing the derived completion described in this paper. Although this is our main application, we hope and expect that our constructions will find use in other contexts.

At least three other notions of derived completion have appeared in the literature, in [11], [8] and [3]. The goal of this paper is to record a precise version of the construction which we can use in future work, but we do make some comments on the relationship of the present construction with those in [11] and [8].

The outline of the paper is as follows. Section 2 develops the preliminary technical material we will require, including material on the theory of S-algebras and module spectra, as well as the theory of Barr and Beck on simplicial and cosimplicial approximations of spectra using "triple" or "monads". Section 3 then defines our derived completion, and develops its elementary properties. In Section 4, we study the behavior of this construction on discrete rings, and show that it coincides with ordinary completion for finitely generated modules over a Noetherian ring. In Section 5, we compare our construction with two other completions, those constructed in [11] and [8]. Section 6 proves our invariance result for S-algebra homomorphisms inducing an isomorphism on π_0, and Section 7 constructs the spectral sequence discussed above. Finally, in Section 8, we develop some interesting examples.

PRELIMINARIES

S-algebras and Module Spectra

As mentioned in the introduction, there are a number of constructions of categories of spectra which admit a coherently associative and commutative smash product. One consequence of these constructions is that one can develop a theory of "ring spectra" as the category of monoid objects in a category of spectra relative to the smash product. As mentioned in the introduction, we elect to work with S-algebras as our notion of ring spectrum. In a similar way, one can define module spectra M over an S-algebra R as spectra equipped with a map $R \wedge M \rightarrow M$ so that the standard algebraic diagrams commute. One can also define the notion of a commutative S-algebra, in terms of the commutativity of an obvious diagram. The commutative S-algebras form a category in their own right, we will denote it by Alg_S. More generally, if A is a commutative S-algebra, we can also construct the category Alg_A of commutative A-algebra spectra. Given any commutative S-algebra A, one can also define a category Mod_A of module spectra over A. Note that because A is commutative, we do not have to specify whether the module is a right or left module. For any commutative S-algebra there exist relative notions of smash products (analogous to tensor products over a ring) and Hom-spectra (analogous to Hom-modules in algebra). See [10] or [12] for the particulars of these theories. We will henceforth work with the S-algebra version of this theory as constructed in [10]. The following proposition summarizes the properties of the categories Alg_A and Mod_A which we will need. The results can all be found in [10], pp. 140–148.

Proposition 2.1: For any commutative S-algebra A, the categories Alg_A and Mod_A can both be equipped with the structure of a Quillen model category (see [7]) with the following properties.

1. A morphism in Alg_A or Mod_A is a weak equivalence if and only if its underlying map of spectra is one.
2. All objects in Alg_A or Mod_A are fibrant.
3. In each category, there is a functorial way to replace each morphism f with a decomposition $f = p \circ i$ with p being a fibration and i, a cofibration which is also a weak equivalence. Similarly for p, a fibration and a weak equivalence and i, a cofibration. In particular, there exist functorial cofibrant and fibrant replacements in Mod_A and Alg_A.
4. For any homomorphism $A \to B$ of commutative S-algebras, the functor $M \to M_A^{\wedge} N$ is a triple in the sense of [2]. See Section 2.3 below.
5. There exists a triple Sym_s on the category Mod_A so that a commutative A-algebra spectrum is precisely the same thing as an algebra over the triple Sym_s, in the sense of [2].

In particular, the category of spectra is equivalent to the category Mod_S, where S is the sphere spectrum, and so the results apply to the category of spectra.

The smash products and Hom constructions are homotopy invariant when the argument modules are cofibrant and/ or fibrant, in the sense made precise in the following proposition.

Proposition 2.2: Let A be a commutative S-algebra. Let $f: M \to M'$ be a weak equivalence of cofibrant A-module spectra, and let N be any fibrant A-module spectrum. Then the induced maps

$$f \underset{A}{\wedge} id_N : M \underset{A}{\wedge} N \to M' \underset{A}{\wedge} N$$

and

$$\mathrm{Hom}_A(f, N) : \mathrm{Hom}_A(M', N) \to \mathrm{Hom}_A(M, N)$$

are both equivalences. Also, if M is cofibrant, and $g: N \to N'$ is a weak equivalence, then the natural map

$$\mathrm{Hom}_A(M, g) : \mathrm{Hom}_A(M, N) \to \mathrm{Hom}_A(M, N')$$

is a weak equivalence.

The smash product and Hom constructions also behave well on cofibrations of module spectra, in the following sense.

Proposition 2.3: Let A be a commutative S-algebra, and let $f: M \to M'$ be a cofibration of A-module spectra, and let N be an A-module spectrum. The induced maps $f\underset{A}{\wedge}id_n$ and $\mathrm{Hom}_A(f,n)$ are cofibrations and fibrations, respectively.

These results allow one to develop spectral sequences as computational tools.

Proposition 2.4: Suppose A is a commutative S-algebra, M is a cofibrant A-module spectrum, and Nis an A-module spectrum. Then there is a spectral sequence with $E_2^{pq} = \text{Tor}_{pq}^{\pi_*(A)}(\pi_*(M), \pi_*(N))$, converging to

$$\to \text{Tot}(\text{Tfib}(M))$$

$\pi_{p+q}(M \underset{A}{\wedge} N)$. The superscript p refers to homological degree, and q refers to internal degree.

Corollary 2.5: Suppose A is a (−1)-connected commutative S-algebra, M is an s-connected cofibrant A-module spectrum, and N is a t-connected A-module spectrum. Then $M \underset{A}{\wedge} N$ is (s+t+1)-connected.

Cosimplicial S-algebras and Module Spectra

Let A be a commutative S-algebra, and let Mod_A denote the category of module spectra over A. Since Mod_A is a Quillen model category the usual notions of homotopy colimits, homotopy limits, the total A-module spectrum of a cosimplicial object in Mod_A as well as its finite stage approximations Tot, and fibrancy of a cosimplicial A-module spectrum all make sense, and they share the properties of the corresponding notions in the category of simplicial sets and the category of spectra. Every cosimplicial object in Mod_A is functorially equivalent to a fibrant one, and we write $(-)_{\text{fib}}$ for this functor. We recall some of the important properties.

Proposition 2.6: Let $F: \Delta \to \text{Mod}_A$ denote a fibrant cosimplicial object in Mod_A. Then there is a canonical natural equivalence

$$\text{Tot}(F) \to \underset{\Delta}{\overleftarrow{\text{holim}}}\, F.$$

Moreover, if we let $\Delta^{(n)}$ denote the full subcategory on the objects of cardinality less than or equal to n+1, then there is a natural equivalence

$$\text{Tot}^n(F) \to \underset{\Delta^{(n)}}{\overleftarrow{\text{holim}}}\, F.$$

Proposition 2.7: Let $F : \underline{C} \times \underline{D} \to \mathrm{Mod}_A$ be a functor, where $\underline{C}$ and $\underline{D}$ are small categories. Then for any object $c \in \underline{C}$, we may construct the homotopy inverse limit object $F_c = \underset{\underline{C} \times \underline{D}}{\mathrm{holim}} F \mid c \times \underline{D}$.

The construction $c \to F_c$ is functorial in c, and we obtain a natural equivalence

$$\underset{\acute{C} \times D}{\mathrm{holim}} \; F \cong \underset{\acute{C}}{\mathrm{holim}} \; F_c.$$

Proposition 2.8: Suppose we have a sequence $X^{\cdot} \to Y^{\cdot} \to Z^{\cdot}$ of fibrant cosimplicial objects in Mod_A, which is levelwise a cofibration sequence. Then the sequences

$$\mathrm{Tot}^n \; X^{\cdot} \to \mathrm{Tot}^n \; Y^{\cdot} \to \mathrm{Tot}^n \; Z^{\cdot}$$

and

$$\mathrm{Tot} \; X^{\cdot} \to \mathrm{Tot} \; Y^{\cdot} \to \mathrm{Tot} \; Z^{\cdot}$$

are cofibration sequences up to homotopy in Mod_A.

Proposition 2.9: Let $X^{\cdot}$ be a fibrant cosimplicial object in Mod_A, and let M denote an object in Mod_A. Then we may form the new cosimplicial object $M \underset{A}{\wedge} N$, and there is a natural equivalence in Mod_A

$$M \underset{A}{\wedge} \mathrm{Tot}^n \; X^{\cdot} \cong \mathrm{Tot}^n ((M \underset{A}{\wedge} X^{\cdot})_{\mathrm{fib}}).$$

Proof.
Follows easily from (see [10] or [12]) the fact that forming smash products over A with a fixed module commutes with homotopy pullbacks of diagrams of the form $X \to Y \leftarrow Z$. We omit the details.

It will also be useful to use the idea of an augmented cosimplicial object in a category. Let Δ_{aug} denote the category obtained by adjoining to Δ the single object -1, which represents the empty set, regarded as a totally ordered set. This description also makes it clear how to define the morphisms in Δ_{aug}. By an augmented cosimplicial object in a category $\underline{C}$, we mean a covariant functor from Δ_{aug} with values in $\underline{C}$. If we have any abelian group-valued functor A on $\underline{C}$, we obtain from any cosimplicial object in $\underline{C}$ an augmented cosimplicial abelian group. As with cosimplicial abelian groups, we may then associate a cochain complex (starting in codimension -1) to this cosimplicial abelian group. We summarize the properties of augmented cosimplicial spectra.

Proposition 2.10: Suppose we are given an augmented cosimplicial object $X^{\cdot}$ in a category $\underline{C}$, and let $\rho X^{\cdot}$ denote the cosimplicial object obtained by restriction to $\Delta \subseteq \Delta_{aug}$. Suppose also that $\underline{C}$ is complete, hence admits a notion of "total object" Tot $\rho X^{\cdot}$. Then the coface map $\Delta^0 : X^{-1} \to X^0$ induces a morphism $\theta : X^{-1} \to \mathrm{Tot}\,(rX^{\cdot})$.

The following result gives a criterion which guarantees that the map η is a weak equivalence, when the underlying category $\underline{C}$ is the category Mod_A.

Proposition 2.11: Suppose $X^{\cdot}$ is an augmented cosimplicial object in the model category Mod_A and that the cosimplicial object $\rho X^{\cdot}$ is a fibrant cosimplicial object in Mod_A. For each t, we obtain an augmented cosimplicial abelian group $\pi_t(X)$, to which we associate a cochain complex $C^*(t)$ as above. If each of the cochain complexes $C^*(t)$ has trivial cohomology, then the map $\theta : X^{-1} \to \mathrm{Tot}\,(rX^{\cdot})$ is a weak equivalence in Mod_A.

Proof
Straightforward verification using the homotopy spectral sequence of a cosimplicial space, see [5], Chapter X.

Finally, we will require a comparison theorem for bicosimplicial A-modules.

Proposition 2.12: Suppose that we have a map of bicosimplicial A-modules $f^{..}:X^{..}\rightarrow Y^{..}$. Suppose that for each $p\geq 0$, the map Tot (X^p_{fib}) $\rightarrow$ Tot (Y^p_{fib}) is a weak equivalence. Then the natural map Tot $(\Delta(X^{..})_{fib})$ $\rightarrow$ Tot $(\Delta(Y^{..})_{fib})$ is a weak equivalence, where Δ denotes the diagonal cosimplicial space. Similarly if we study the level wise simplicial objects obtained by holding q fixed.

Proof
Straightforward, and we omit it.

The Theory of Barr and Beck

Our theory of completion will make use of the theory of the cosimplicial object of a triple, a notion discussed by Barr and Beck [2]. We will need to use some of the comparison theorems from that paper, so we review that theory here. For the definition of a triple and algebras over a triple, see [19].

Suppose a category $\underline{C}$ is equipped with a triple T, and $c \in \underline{C}$. Then we may define the cosimplicial resolution of c relative to T to be the cosimplicial object $T^{.}(c)$ defined by $T^k(c) = T^{k+1}(c)$, and where the cofaces and codegeneracies are defined by $\Delta^s:T^{k+1}(c)\rightarrow T^{k+2}(c)=T^s(\eta(T^{k+1-s}(c)))$

and

$\sigma^t:T^{k+1}(c)\rightarrow T^k(c)=T^t(\mu(T^{k-t-1}(c)))$.

Note that $T^{.}(c)$ extends canonically to an augmented cosimplicial object $T^{.}_{aug}(c)$ by setting $T^{-1}_{aug} = c$, and letting $\delta^0 : T^{-1}_{aug} \rightarrow T^0_{aug}$ be the canonical inclusion $\eta:c\rightarrow T(c)$. This means that we have a natural map $\theta:c \rightarrow$ Tot $(T^{.}(c))$.

Suppose now that the category $\underline{C}$ is Mod_A. For any A-module spectrum M, will denote by $\mathsf{T}_{\mathrm{fib}}^{\cdot}(M)$ the functorial fibrant replacement for $\mathsf{T}^{\cdot}(M)$. We will need criteria which guarantee that the map $q_{\mathrm{fib}}(M){:}M \to \mathrm{Tot}(\mathsf{T}_{\mathrm{fib}}^{\cdot}(M))$ is a weak equivalence in Mod_A, where $q_{\mathrm{fib}}(M)$ is the composite $M \overset{\theta}{\to} \mathrm{Tot}(\mathsf{T}(M)) \to \mathrm{Tot}(\mathrm{Tfib}(M))$. Barr and Beck now prove the following result.

Theorem 2.13: Suppose that M is equipped with a T-algebra structure. Then $q_{\mathrm{fib}}(M)$ is a weak equivalence.

We will derive a useful corollary of this result, using the following lemma.

Lemma 2.14: Let S and T be two triples on the model category Mod_A, and suppose we are given a natural transformation $v{:}\ S \to T$ of triples. We may construct the bicosimplicial A-module C given by $C^{pq} = \mathsf{T}_{\mathrm{fib}}^p (S_{\mathrm{fib}}^q (M))$. We may also regard the cosimplicial spectrum $\mathsf{T}_{\mathrm{fib}}^{\cdot}$ as a bicosimplicial spectrum, constant in the q-direction, and denote it by $\mathsf{T}_0^{\cdot \cdot}$. Then the evident bisimplicial map $qT{:}\ \mathsf{T}_0^{\cdot \cdot} \to C^{\cdot \cdot}$ Induces an equivalence on total A-modules.

Proof
By applying Proposition 2.12, it will suffice to prove that each of the natural maps

$$T^p(M) \to \underset{q}{\mathrm{Tot}}(T^p S^q(M)_{\mathrm{fib}})$$

is a weak equivalence of A-module spectra. By 2.11, we need only verify that the cohomology of the cochain complex A^* attached to the augmented cosimplicial group $\pi_i(T^p((S_{\mathrm{fib}}(M)_{\mathrm{aug}})))$ is trivial. However, for each $q \geq 0$, we have the map $h^q{:}T^p(S^q(M)) \to T^p(S^{q-1}(M))$ given by the composite

$$T^p(S^q(M)) \xrightarrow{T^p(v(S^{q-1}(M)))} T^p(T S^{q-1}(M)) \xrightarrow{T^{p-1}(\mu^T(S^{q-1}(M)))} T^p(S^{q-1}(M)).$$

As in the proof in [2] of Theorem 2.13, the operators $\pi_1(h^q)$ yield a contracting homotopy for the cochain complex A^*, which gives the result.

We also obtain the following.

Proposition 2.15: Let S and T be two triples on the model category Mod_A, and suppose we are given a natural transformations $v: S \to T$ and $\mu: T \to S$ of triples. Then the natural map

$$\mathrm{Tot}\, \mathcal{S}_{\mathrm{fib}}(M) \to \mathrm{Tot}\, \mathcal{T}_{\mathrm{fib}}(M)$$

is an equivalence, where S and T denote the cosimplicial resolutions of M associated with S and T, respectively.

Proof
This follows directly from Theorem 5.3 in [16]. It is also a straightforward consequence of the work in [2] or [4].

We will also find it useful to discuss the simplicial resolution of a T-algebra X, for any triple T.

Definition 2.16
Let T be a triple on a category $\underline{C}$, and let X be any T-algebra, with structure morphism $\alpha: T\, X \to X$. The simplicial resolution of X relative to T is the simplicial object $T_.(X)$, defined on objects by

$$T_k(X) = \underbrace{T \circ T \circ \cdots \circ T}_{k+1 \text{ factors}}(X) \quad \text{for } k \geq 0$$

and on morphisms by

$$\begin{cases} d_i = T^i \mu(T^{k-i-1}(X)) & \text{for } 0 \leq i \leq k-1 \\ d_k = T^k(\alpha) \\ s_i = T^{i+1}\eta(T^{k-i}(X)) & \text{for } 0 \leq i \leq k. \end{cases}$$

The structure map α gives a map $\varepsilon(X):|T.(X)|\to X$, where X denotes the constant simplicial object with value X.

Proposition 2.17: Let T denote a triple on the category of spectra. Then for any T-algebra spectrum X, the map $\varepsilon(X)$ induces an isomorphism on homotopy groups.

Proof

The structure map $\alpha: TX \to X$ provides an extension of the functor T. to the larger category Δ^{op}_{aug}, which we will denote by T^{aug}. We will refer to a functor from Δ^{op}_{aug} to a category $\underline{C}$ as an augmented simplicial object in $\underline{C}$. For any augmented simplicial abelian group A·, we may construct the associated chain complex $C_*(A·)$, with the alternating sum of face maps as boundary operator. For any augmented spectrum X·, we denote by $\rho(X·)$ its restriction to the subcategory $\Delta^{op} \subseteq \Delta^{op}_{aug}$, so $\rho(X·)$ is a simplicial spectrum. The augmented spectrum gives rise to a map $v(X·):\rho(X·)\to X-1$· by analogy with the cosimplicial case, it is now easy to verify that given any augmented simplicial spectrum X·, if it is the case that for each i, the chain complex $C_*\pi s(X·)$ has trivial homology, then the map $v(X·)$ is a weak equivalence. For any triple T on the category of spectra, and every integer t, the complex $C_*(\rho_t T^{aug}(X))$ has trivial homology, since the operators

$$\pi_t \eta(T^{k+1}(X)) : \pi_t T_k^{aug}(X) \to \pi_t T_{k+1}^{aug}(X)$$

provide a contracting homotopy for it. Note that the operator even makes sense for k=−1, with $T^0(X)=X$. This gives the result.

DEFINITIONS

Let A denote a commutative S-algebra and B a commutative A-algebra spectrum. The construction in [10] and [12] of smash products over A make the functor $T_A(-; B) = B \wedge_A -$ into a triple on the category Mod_A. Consequently, we may construct the cosimplicial resolution of $T_A(-; B)$ as a functor on the category of A-module spectra to the category of co-simplicial A-module spectra. We will write $\tilde{T_A}(-; B)$ for the functorial fibrant replacement of this cosimplicial object in Mod_A.

We recall from [18] that for any commutative S-algebra A, there is a closed model structure on the category of A-algebras. Moreover, one may functorially replace any A-algebra by a weakly equivalent A-algebra which is cofibrant.

Definition 3.1
Let A denote a commutative S-algebra, B a commutative A-algebra, and M an A-module. We define the derived completion of M at the A-algebra B to be the total spectrum of the cosimplicial spectrum $\tilde{T_A}(M; \tilde{B})$, where $\tilde{B}$ is the cofibrant replacement for B in the category of commutative A-algebras. We write $M_{\tilde{B}}^{\wedge}$ for Tot $(\tilde{T_A}(M; B))$.

The following propositions summarize the most important properties of this construction.

Proposition 3.2: Let $A \to B \to C$ be a diagram of commutative S-algebras. Then the following statements all hold.

1. The construction $M \to M_B^{\wedge}$ is functorial for homomorphisms of A-modules. The map on B-completions induced by a homomorphism $f: M \to N$ of A-modules will be denoted by $f_B^{\wedge}$.
2. Let $f: M \to N$ be a homomorphism of A-modules, where A is a commutative S-algebra, and let B denote a commutative A-algebra. Suppose that $id_B \wedge_A f$ is a weak equivalence of B-module spectra.

Then the map on completions $f_{\hat{B}}$ is also a weak equivalence of spectra.

3. Let $M \to N \to P$ be a cofibration sequence of A-module spectra, where A is a commutative S-algebra. Let B be a commutative A-algebra spectrum. Then the sequence $M_{\hat{B}} \to N_{\hat{B}} \to P_{\hat{B}}$ is a cofibration sequence up to homotopy.

4. Let $A \to B \to C$ be a diagram of commutative S-algebras, and let M denote a B-module spectrum. M may be regarded as an A-module spectrum M^A. Suppose that the natural map $C \wedge_A M^A \to C \wedge_B M$ is a weak equivalence of spectra. Then the natural map $(M^A)\hat{C} \to M \hat{C}$ is also an equivalence of spectra.

5. There is a natural transformation $\eta: M \to M_{\wedge}$.

6. Suppose $A \to B$ is a homomorphism of commutative S-algebras. Suppose M is an A-module spectrum, for which the A-module structure admits a B-module structure extending the given A-module structure. Then the natural map $\eta: M \to M\hat{B}$ is an equivalence of spectra.

Proof

(1) is immediate from the constructions. (2) is an immediate consequence of the standard fact that a levelwise equivalence of fibrant spectra induces a weak equivalence of Total spectra. (3) and (4) follow from the fact that smash products with a fixed module preserve homotopy cofibration sequences, and that the total spectrum carries levelwise homotopy cofibration sequences to homotopy cofibration sequences. (5) follows from 2.13. (6) is a general property of the cosimplicial resolution of a triple.

Remark 3.1

We note that although we define the derived completion of an A-module spectrum M at a homomorphism $f: A \to B$ of commutative S-algebras to be the total spectrum of $T_A^\cdot$ $(M; B)$, it is useful to recall that it is therefore the total space of the tower of fibrations

$$\cdots \to \operatorname{Tot}^n(T_A^\cdot(M; B)) \to \operatorname{Tot}^{n-1}(T_A^\cdot(M; B)) \to \cdots \to \operatorname{Tot}^1(T_A^\cdot(M; B)) \to \operatorname{Tot}^0(T_A^\cdot(M; B))$$

and that the tower of fibrations is actually functorial as well. This tower of fibrations gives filtrations on the homotopy groups and other invariants, and will likely be interesting in future K-theoretic applications of these ideas.

THE CASE OF RINGS

For any ring A, we may construct the Eilenberg–MacLane spectrum (A). It is direct from the constructions of [10] and [12] that (A) is in a canonical way a S-algebras, and that if the ring is commutative, (A) is a commutative S-algebra. Left and right modules over the ring A yield, via the construction, left and right module spectra over the S-algebra A. The goal of this section is to analyze how the derived completion construction given in the preceding section applies to S-algebras obtained via this construction. Specifically, we will show that it coincides with ordinary completion for finitely generated modules over commutative Noetherian rings, in the sense, $\mathbb{H}(M)^\wedge_{\mathbb{H}(A/I)} \cong \mathbb{H}(M^\wedge_I)$ for any finitely generated A-module M and any ideal $I \subseteq A$, and where

$M^\wedge_I$ denotes the usual algebraic completion operation at the ideal I. Note that for a commutative ring A, any left A-module can be canonically regarded as a right A-module as well, so we will simply refer to A-modules without specifying right or left modules.

Proposition 4.1: Let A be a ring, M and N be A-modules. Let $\overline{\mathbb{H}(M)}$ denote a cofibrant replacement of the (A)-module spectrum (M). Then

$$\pi_*(\overline{\mathbb{H}(M)} \wedge_{\mathbb{H}(A)} \mathbb{H}(N)) \cong \operatorname{Tor}_*^A(M, N)$$

as graded groups. $M \hat{\wedge}_A N$ Denotes the spectrum level construction of "smash product over A", and Tor denotes the usual algebraic derived functors.

Proof

This results follows from the Künneth spectral sequence whose E_2-term is $\text{Tor}_*^A (M, N)$, and which collapses for dimensional reasons.

This means that for any family of k left A-modules $M_1, M_2, \dots, M_k$, we may define groups

$$\text{MultiTor}_i^A(M_1, M_2, \dots, M_k)$$

as the ith homology of the complex $R(M_1) \otimes_A R(M_2) \otimes_A \otimes_A R(M_k)$ where $R(M_j)$ denotes an A-projective resolution òf M_j.

Proposition 4.2: Let A be a commutative S-algebra, and let M_1, $M_2, \dots$, M_k denote a family of left A-modules. The Künneth spectral sequence generalizes to a spectral sequence with E_2-term

$$\text{MultiTor}_*^A(M_1, M_2, \dots, M_k)$$

converging to

$$\pi_*(M_1 \wedge_A M_2 \wedge \cdots \wedge_A M_k).$$

Proposition 4.3: Let A be a commutative ring, and let $B = A/I$, where I is an ideal in A. Suppose further that M is a B-module, which we regard as an A-module by restriction of scalars. Then the natural map $M \to M \hat{_B}$ is an equivalence.

Proof: This is immediate from Proposition 3.2, (6).

What we have now shown is that for any B-module M, the derived completion M $\hat{}_{\hat{B}}$ is equivalent to M itself, regarded as a module spectrum over A. We wish to extend this to a result valid for all finitely generated A-modules over a Noetherian ring. We recall first that the definition of the I-adic completion of an A-module M is

$$\lim_{\overleftarrow{k}} M/I^k M.$$

We may also consider the homotopy inverse limit of the pro-A-module $\{M/I^k M\}_{k\geq 0}$. Recall from [5] that for any inverse system of spectra $\{X_k\}_{k\geq 0}$, there is a short exact sequence

$$0 \to \lim^1\{\pi_{i-1} X_k\}_{k\geq 0} \to \pi_i \operatorname*{holim}_{\overleftarrow{k}} X_k \to \lim_{\overleftarrow{k}} \pi_i X_k \to 0.$$

In our case, the inverse system $\{\pi_i - 1 X_k\}_{k\geq 0}$ consists entirely of surjective maps, it is a standard result that its $\lim^1$-term vanishes, leaving us with the isomorphism

$$\pi_i(\operatorname*{holim}_{\overleftarrow{k}} M/I^k M) \cong \lim_{\overleftarrow{k}} M/I^k M \quad \text{for } i = 0$$

and

$$\pi_i(\operatorname*{holim}_{\overleftarrow{k}} M/I^k M) \cong 0 \text{ for } i \neq 0.$$

In other words

$$\operatorname*{holim}_{\overleftarrow{k}} M/I^k M \cong \mathbb{H}(M_I^\wedge).$$

We now wish to show that

$$M_B^\wedge \cong \underset{k}{\mathrm{holim}}\, M/I^k M.$$

Consider the inverse system of cosimplicial A-module spectra $\{T_A^\cdot(M/I^kM;B)\}_{k\geq 0}$. We have the natural map

$$\theta : T_A^\cdot(M; B) \to \{T_A^\cdot(M/I^k M; B)\}_{k\geq 0}$$

where $T_A^\cdot(M;B)$ is considered as a constant pro-cosimplicial A-module spectrum. By taking total spectra and homotopy inverse limits (in the k-direction), we obtain a map of an A-module spectra $M\hat{B} \to \underset{k}{\mathrm{holim}}\, \mathrm{Tot}\, T_A^\cdot(M/I^k M; B)$. By 4.3, we have that $\mathrm{Tot}\, T_A^\cdot(M/I^k M; B) \cong M/I^k M$, so we obtain a natural map

$$\lambda : M_B^\wedge \to \underset{k}{\mathrm{holim}}\, M/I^k M \cong \mathbb{H}(M_I^\wedge).$$

Our result is now

Theorem 4.4
For a Noetherian commutative ring A, and a finitely generated A-module M, the map $\lambda : M\hat{B} \to \mathbb{H}(M_I)$ described above is an equivalence.

Proof
It is standard that homotopy inverse limits of cosimplicial spectra commute with taking total spectra, so it is enough M_I to verify that the maps

$$T_A^i(M; B) \to \underset{k}{\mathrm{holim}}\, T_A^i(M/I^k M; B)$$

are equivalences for each i. In order to prove this, it will suffice to show that the pro-A-module $\{ T_A^i (M / I^k M; B) \}_{k \geq 0}$ is isomorphic, as pro-abelian groups, to the constant pro-abelian group with value T_A^i (M; B). By 4.2, this means that it will suffice to show that the pro-abelian group

$$\{\operatorname{MultiTor}_*^A (\underbrace{B, B, \ldots, B}_{s \text{ factors}}, M / I^k M)\}_{k \geq 0}$$

is isomorphic to the constant pro-abelian group with value

$$\operatorname{MultiTor}_*^A (\underbrace{B, B, \ldots, B}_{s \text{ factors}}, M).$$

For any A-module M, we let $\underline{I}M$ denote the pro-A-module $\{I^k M\}_{k \geq 0}$. Recall (see [1]) that the category of pro-A-modules is itself an abelian category, and so the notion of exact sequence has meaning. The construction $M \rightarrow \underline{I}M$ gives a functor from the category of finitely generated A-modules to the category of pro-A-modules, which is exact in the sense that it carries exact sequences to exact sequences. The exactness follows from the usual proof of exactness of completion for finitely generated modules over a Noetherian ring (see [9]). This same exactness also shows that for complexes of finitely generated A-modules, homology commutes with the functor $\underline{I}$, so that we have

$$\operatorname{MultiTor}_*^A (\underbrace{B, B, \ldots, B}_{s \text{ factors}}, \underline{I}M) \cong \underline{I}\operatorname{MultiTor}_*^A (\underbrace{B, B, \ldots, B}_{s \text{ factors}}, M).$$

It is now clear that $\operatorname{MutiTor}_*^A \left(\underbrace{B, B, \ldots \ldots B}_{s \text{ factors}}, M \right)$ is a B-module, and therefore I acts trivially on it, which means that the pro-abelian group $\underline{I}$

$\mathrm{MutiTor}_*^A\left(\underbrace{B,B,\ldots\ldots\ldots B}_{s\,\text{factors}},M\right)$ is pro-isomorphic to the zero group. We have an exact sequence of pro-abelian groups

$$\underline{I}M \rightarrow M \rightarrow M/\underline{I}M$$

so we may conclude that the map $M \rightarrow M/\underline{I}M$ induces an isomorphism on homology as pro-groups. This is the required result.

COMPARISON WITH OTHER CONSTRUCTIONS

In this section we will prove some results comparing the present construction with those of Greenlees and May [11] and Dwyer, Greenlees and Iyengar [8].

The Greenlees–may completion

This completion construction considers the situation of a commutative ring A and a finitely generated ideal I. It begins with a purely algebraic construction of derived functors of completion, and then produces a spectrum level construction. We will compare the derived functors defined in [11] with the homotopy groups of our derived completion, and proved that they agree. It appears likely that the methods of Section 7 would allow a comparison for the spectrum level construction, but we do not carry out that here. We will write $B=A/I$, and let S denote any finite generating set for I. In [11], the authors construct a chain

complex C_*^s, essentially a colimit of Koszul complexes based on S, and define their derived completion of an A-module M, which we will denote by M_s^{gm}, by

$$M_S^{gm} = \mathrm{Hom}_A(C_*^S, M).$$

The construction depends on S, but there are canonical isomorphisms between the homology groups constructed using different finite generating sets S. These are the derived functors of completion at I applied to the module M. There is a natural chain map $C_*^S \xrightarrow{\gamma} A_*$, where A_* denotes the chain complex, concentrated in degree 0, with zeroth module equal to A. The map γ now induces a natural map

$$\xi : M \to M_S^{gm}$$

where M denotes the module M regarded as a chain complex concentrated in degree zero. We now have the following.

Lemma 5.1: Let M be any B-module, regarded as an A-module by restriction of scalars. Then the natural map $\xi : M \to M_S^{gm}$ is a quasi-isomorphism of chain complexes

Proof
Immediate consequence of Lemma 1.3 of [11].

It is also immediate from the definitions that the construction $M \to M_S^{gm}$ extends to chain complexes of A-modules. The following is now easy to prove.

Lemma 5.2: The construction $C_* \to (C_*)_S^{gm}$ preserves quasi-isomorphisms. Given a chain complex C_* which is bounded below, and so that the A-module structure on $H_i(C_*)$ extends to a B-module structure for all i, then the natural map $\xi : C_* \to (C_*)_S^{gm}$ is a quasi-isomorphism.

Proof
The first statement is straightforward from the definitions. The second one follows from the existence of a "Postnikov tower" for any bound-

ed below chain complex, where the relative terms are resolutions of the A-modules $H_i(C_*)$. The first statement implies that if $R_*(M)$ is a resolution of an A-module M, then the map $R_*(M) \to M$ induces an equivalence $(R_*(M))_s^{gm} \to M_s^{gm}$. The result now follows from Lemma 5.1 above, together with the evident fact that the construction carries exact sequences of chain complexes to exact sequences.

In order to compare M_s^{gm} with $M_\wedge^B$, we will need to reinterpret $M_\wedge^B$ as a chain complex. We first begin by constructing a cofibrant version B_* of B. This can be done, for example, by using the free commutative A-algebra functor on sets. This is a triple on the category of sets, and for any A-algebra yields a simplicial resolution of B, which is levelwise free as an A-module. It is also a simplicial commutative ring. The cosimplicial spectrum $T_A (M; B_*)$ which constructs $M_\wedge^B$ is now a cosimplicial simplicial abelian group, and its total space is a simplicial abelian group. As such, it corresponds to a chain complex. On the other hand, one can define the notion of the chain complex associated with a spectrum. One begins with the associated chain complex functor C_* for the notion of spaces one is dealing with, the singular complex in the case of topological spaces, and the simplicial chains for simplicial sets. For any spectrum $X = \{X_i\}_{i \geq 0}$, with structure maps $\sigma_i : \Sigma X_i \to X_{i+1}$, one defines $ch(X)$ to be the colimit of the system

$$\cdots \to \Sigma^{-i} C_*(X_i) \to \Sigma^{-(i+1)} C_*(X_{i+1}) \to \Sigma^{-(i+2)} C_*(X_{i+2}) \to \cdots.$$

Now, the cosimplicial spectrum corresponds to a cosimplicial chain complex, by applying ch levelwise. We write $E(k, M)_*$ for the chain complex in codimension k. It is defined by

$$E(k, M)_* = ch\left(\bigwedge_k B \wedge M\right).$$

For any cosimplicial chain complex $C(-)_{*}$, we define a chain complex $DC(-)_*$ by

$$\mathcal{D}C(-)_* = \prod_k \Sigma^{-k} C(k)_*$$

on the level of graded groups (Σ^t just denotes a shift of degrees by t), and the boundary map comes from the bicomplex structure on this graded group, with one boundary being the one existing on each of the complexes $\Sigma^{-k}C(k)_*$, and the other being the alternating sum of the coface maps.

Proposition 5.3: Let $k \to A(k)_*$ denote any cosimplicial simplicial abelian group. Then there are canonical isomorphisms

$$\pi_s(\mathrm{Tot}_k A(k)_*) \cong H_s(\mathcal{D}ch(A(-)_*))$$

for all s.

Proof: This is standard. See for example [5].

We now write $\mathcal{E}_*(M)$ for the chain complex $DE(-,M)_*$. The construction $\mathcal{E}_*$ is clearly functorial in M, and $H_*(\mathcal{E}_*(M))$ is canonically isomorphic to $\pi_*(M^\wedge)$. We note that $\mathcal{E}_*$ is a chain complex of A-modules, and that the standard map from the constant cosimplicial A-module with value M to $T_A(M; B_*)$ induces a natural transformation $\theta : M_* \to \mathcal{E}_*(M)$, where M_* denotes M regarded as a chain complex concentrated in degree zero.

We wish to compare the groups $H_*(M)_s^{gm}$ and $H_*(\mathcal{E}_*(M))$. The idea will be to compare each of the complexes in question with the complex $\mathcal{E}_*(M)_s^{gm}$. The natural transformation θ produces a natural map $\theta : M_s^{gm} \to \mathcal{E}_*(M)_s^{gm}$ and the transformation ξ above yields a natural map $E : \mathcal{E}_*(M) \to \mathcal{E}_*(M)_s^{gm}$. We now have the following result.

Proposition 5.4: The natural maps Θ and E are quasi-isomorphisms. Therefore the groups $H_*(M)_s^{gm}$ and $H_*(\mathcal{E}_*(M))$ are canonically isomorphic.

Proof: We begin with E. We note first that $\mathcal{E}_*(M)$ is itself obtained as the total complex of a bicomplex, and can be filtered by the codimension k. Note that this is a decreasing filtration. The subquotients are the complexes $ch(\bigwedge_k B \wedge M)$. We obtain a filtration on $\mathcal{E}_*(M)_s^{gm}$ which is compatible with the filtration on $\mathcal{E}_*(M)$ under the natural map Ξ. Both complexes are obtained as inverse limits of the quotients by the terms in the filtrations, so in order to prove the result it will suffice to show that the natural map

$$ch\left(\bigwedge_k B \wedge M\right) \to \left(ch\left(\bigwedge_k B \wedge M\right)\right)_S^{gm}$$

is a quasi-isomorphism for all k. The reduction uses the $\lim^1$ sequences for the homology of inverse limits of chain complexes. But now, in view of Lemma 5.2 above, it will suffice to show that the A-module structure on $H_i(ch(\bigwedge_k B \wedge M))$ extends to a B-module structure. But this is clear, since the homology of these complexes are the MulitTor groups of M with coefficients in B, and the elements of B act on B by multiplication. This shows that Ξ is a quasi-isomorphism. For Θ, we note that the Greenlees–May complex C_*^s is a colimit of subcomplexes $C_*^s(k)$, which are quasi-isomorphic to the Koszul complexes $K(k)$ based on the set $\{s^k\}_{s \in S}$. It follows that $(M)_S^{gm}$ is expressed as the inverse limit of complexes $(M)_S^{gm}(k)$, which are quasi-isomorphic to $\text{Hom}_A(K(k), M)$, and similarly for chain complexes. Therefore, there is also a description of $\mathcal{E}_*(M)_S^{gm}$ as an inverse limit of quotient complexes $\mathcal{E}_*$

$(M)_s^{gm}$ (k), which is compatible (under the map Θ) with the inverse limit

description of $(M)_s^{gm}$. It therefore suffices to show that the natural map

$$\mathrm{Hom}_A(\mathcal{K}(k), M) \to \mathrm{Hom}_A(\mathcal{K}(k), \mathcal{E}_*(M))$$

is a quasi-isomorphism. To see this, we note that there is a canonical isomorphism of complexes $\mathrm{Hom}_A(K (k), \mathcal{E}_* (M)) \to \mathcal{E}_* \mathrm{Hom}_A(K (k), M)$. We also note that the complex $\mathrm{Hom}_A(K (k), M)$ admits a finite Postnikov tower, with the relative quotients being quasi-isomorphic to resolutions of the various A-modules $H_i(\mathrm{Hom}_A(K (k), M))$. This therefore induces a similar tower on $\mathcal{E}_* \mathrm{Hom}_A(K (k), M)$, and it therefore suffices to show that the natural map $R_*(H_i) \to E_*(R_*(H_i))$ is a quasi-isomorphism for all i, where

$$H_i \to H_i(\mathrm{Hom}_A(K (k), M)).$$

Since a resolution of an A-module is quasi-isomorphic to that module regarded as a constant chain complex, we have reduced the problem to verifying that $H_i \to \mathcal{E}_* (H_i)$ is a quasi-isomorphism for all i, or equivalently that $H_i \to (H_i)_{\hat{B}}$ is an equivalence. The action of any element s^k on $K(k)$ is easily seen to be chain null homotopic, and the similar result follows for $P = \mathrm{Hom}_A(K (k), M)$. Each H_i therefore has the property that every element of the form s^k, for $s \in S$, acts by zero on H_i. It follows that there exists an integer l so that the ideal I^l acts trivially on H_i. Now consider the filtration $\{I^s H_i\}$ on H_i. It is a finite filtration, and the subquotients $I^s H_i / I^{s+1} H_i$ are A-modules so that I acts trivially. Using the exactness property of our derived completion construction, it suffices to prove that the natural maps

$$I^s H_i / I^{s+1} H_i \to (I^s H_i / I^{s+1} H_i)_{\hat{B}}$$

are equivalences. But this is Proposition 4.3. The following corollary is now immediate.

Corollary 5.5: Let A be any commutative ring, I any finitely generated ideal, with finite generating set S. Let M be any A-module. Then the Greenlees–May completion M_S^{gm} and the derived completion $M_{A/I}^{\wedge}$ are canonically equivalent in the homotopy category of A-module spectra, where A is regarded as an S-algebra via the Eilenberg–MacLane construction.

The Dwyer–greenlees–Iyengar Completion

In [8], the authors construct a version of completion which is inspired by Morita theory. The idea is as follows. Let A and B be commutative S-algebras, and let $f\colon A\to B$ be a homomorphism of commutative S-algebras. One can then construct an S-algebra $\varepsilon = \varepsilon_A(B) = \mathrm{Hom}_A(B,B) = \mathrm{End}_A(B)$, where the Hom constructions are in the category of A-modules. B now becomes a left E-module. Further, for any A-module M, we may form $B_A^{\wedge}M$, and it becomes a left E-module as well. One now constructs $\mathrm{Hom}_\varepsilon(B,B_A^{\wedge}M)$, and note that there is a natural homomorphism $M\to\mathrm{Hom}_\varepsilon(B,B_A^{\wedge}M)$, whose adjoint is the map

$$B\wedge M \to B \underset{A}{\wedge} M$$

of left E-modules, where the action on $B\wedge M$ is on the first factor. We will call the spectrum $\mathrm{Hom}\varepsilon(B,B\underset{B}{\wedge}M)$, the DGI completion of A at the homomorphism f. The main subject of [8] is duality theory in module spectra, and this notion of completion is best understood in the context where B satisfies certain finiteness conditions, which is where the duality theory of [8] takes place. We will see that in our situation, there is a related (more restrictive) finiteness condition, under which we will prove that the DGI completion agrees with our derived completion.

Given A, B, M, f, and E as above, we construct the cosimplicial A-module spectrum $T_A(B;M)$ as in Section 3. We can now construct the cosimplicial B-module spectrum $B_{\hat{A}} T_A(B;M)$, to obtain a cosimplicial E-module $M^{\cdot}$, and construct the cosimplicial E-module spectrum

$\text{Hom}_{\mathcal{E}}(B, \mathcal{M}^{\cdot})$.

Of course, there is the canonical map $M \to T_A(B;M)$, consequently the induced map $M \to T_A(B;M)$, and therefore a natural map

$\alpha : \text{Hom}_{\mathcal{E}}(B, B \underset{A}{\wedge} M) \to \text{Hom}_{\mathcal{E}}(B, \mathcal{M}^{\cdot})$

where the left hand side is the DGI completion of M.

Proposition 5.6: The map $\text{Tot}(\alpha)$ of total spectra induced by α is an equivalence of spectra.

Proof: One readily checks that there is an equivalence

$\text{Hom}_{\mathcal{E}}(B, \text{Tot}\,\mathcal{M}^{\cdot}) \to \text{Tot}(\text{Hom}_{\mathcal{E}}(B, \mathcal{M}^{\cdot}))$

and reduces to proving that the natural map $B_A^{\wedge} M \to \text{Tot}(M)$ is an equivalence. This follows readily from Theorem 2.13.

Since we have a natural map from an A-module M to its DGI completion, we obtain a natural map

$\rho : T_A(M;B) \to \text{Hom}_{\varepsilon}(B,M)$

and the corresponding map of total spectra from $M_{\wedge}^{B}$ to $\text{Tot}(\text{Hom}_{\varepsilon}(B,M))$, which in the homology category can be interpreted as a map from

$M_{\wedge}^{B}$ to the DGI completion of M, according to Proposition 5.6.

Lemma 5.7: Suppose that B is finitely built from A in the sense of [8], and that the A-module M is also finitely built from A. Then the map ρ defined above is an equivalence. In other words, when B is finitely built from A, our derived completion agrees with the DGI completion.

Proof

We sketch a proof. One can write $\operatorname{Hom}_\varepsilon(B, \operatorname{Tot}M)$ as the total spectrum of the cosimplicial spectrum

$$k \mapsto \operatorname{Hom}_\varepsilon(B, B \wedge_A T_A^k(M; B))$$

and in order to prove the result it will clearly suffice to prove that the natural maps

$$T_A^k(M; B) \to \operatorname{Hom}_\varepsilon(B, B \wedge_A T_A^k(M; B))$$

are equivalences. From the definition of T_A, it will now suffice to show that for any module N which is finitely built from A, the natural map

$$B \wedge_A N \to \operatorname{Hom}_\varepsilon(B, B \wedge_A B \wedge_A N)$$

is an equivalence. This follows from the three facts given below, which are easily obtained using the results of [8].

- The correspondence D defined by $D(N) = \operatorname{Hom}_A(N, A)$ is a contra variant equivalence of categories from the category of modules which are finitely built over A to itself. There is a canonical equivalence $N \cong D^2(N)$.

- Given any A-module N which is finitely built from A, there is a canonical equivalence
$$\operatorname{Hom}_\varepsilon(B, B \wedge_A D(N)) \to \operatorname{Hom}_\varepsilon(B \wedge_A N, B).$$

- For any A-module N which is finitely built from A, there is an equivalence of left E-modules

$$B \underset{A}{\wedge} D(B \underset{A}{\wedge} N) \cong \mathcal{E} \underset{A}{\wedge} D(N)$$

where the E-action on the left hand factor is on the left hand factor B, and on the right hand side by left multiplication on the E-factor.

We now extend this lemma into the main result of this section. We recall the notion of proxy finiteness from [8]. Let A be a commutative S-algebra. The A-module B is said to be proxy finite if there is an A-module K so that K is finitely built from A, and so that K is built from B. The DGI completion uses arbitrary A-modules as input, while our construction requires a commutative A-algebra. Consequently, this notion as it stands is not useful for us. We define the following replacement notion which is needed in our context.

Definition 5.8: Let A be a commutative S-algebra, and let B be a commutative A-algebra, with structure homomorphism f: A→B. We say B is algebra proxy finite over A if there is a commutative diagram

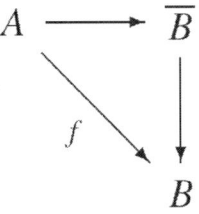

of commutative S-algebras, where the A-module $\bar{B}$ is finitely built from A, and built from B. The following lemma makes this notion useful.

Lemma 5.9: Let

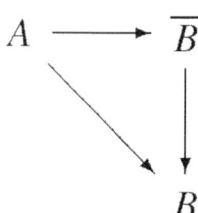

be a commutative diagram of commutative S-algebras as above, and suppose $\overline{B}$ is built from B. Then for any A-module M, the natural map

$$M_{\overline{B}}^{\wedge} \;\to\; M_{B}^{\wedge}$$

is an equivalence of A-module spectra.

Proof: We consider the bicosimplicial spectrum $B^{\cdot\cdot}$ defined by

$$(k, l) \;\mapsto\; T_A^k(T_A^l(M; \overline{B}); B).$$

It is equipped with a natural map

$$\eta : T_A(M; \overline{B}) \to B^{\cdot\cdot}$$

when $T_A(M;\overline{B})$ is regarded as a cosimplicial spectrum constant in the k-direction. We first observe that η induces an equivalence on total spectra. To verify this only required to show that for any $\overline{B}$-module N, the natural map $N \to N_{\overline{B}}^{\wedge}$ is an equivalence. This follows from the fact that the total spectrum construction respects calibration sequences, together with Proposition 3.2, part 5. There is also a natural map $v : T_A(M;B) \to B$, where T_A is regarded as a cosimplicial spectrum constant in the l-direction. It too induces an equivalence on total spectra. It is easy to see that to verify this only required to show that for any A-module N, the natural map

$$B \wedge_A N \;\to\; \mathrm{Tot}(B \wedge_A T_A(N; \overline{B}))$$

is an equivalence of spectra. But, there is an evident equivalence of cosimplicial spectra

$$B \underset{A}{\wedge} T_A^{\cdot}(N; \overline{B}) \simeq T_A^{\cdot}(B \underset{A}{\wedge} N; \overline{B})$$

and we are reduced to showing that the map

$$\eta : B \underset{A}{\wedge} N \to \mathrm{Tot}\, T_A^{\cdot}(B \underset{A}{\wedge} N; \overline{B}) \cong (B \underset{A}{\wedge} N)_{\overline{B}}^{\wedge}$$

is an equivalence. But, since B is an $\overline{B}$-algebra spectrum, it follows directly that $B_A^{\wedge}N$ admits a $\overline{B}$-module structure extending the A-module structure, and the result now follows directly fromProposition 3.2, part 5. It is finally not difficult to check that we have a commutative diagram in the homotopy category

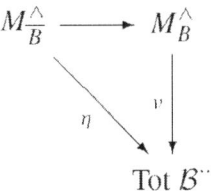

$$M_{\overline{B}}^{\wedge} \longrightarrow M_B^{\wedge}$$

where the horizontal map is the usual map arising from the functoriality of the completion construction. This gives the result.

We now draw our main conclusion.

Theorem 5.10: Let f: A→Bbe a homomorphism of commutative S-algebras, and suppose that B is algebra proxy finite. Then for any A-module Mwhich is finitely built from A, the map ρ defined above from

$M_{\underset{B}{\wedge}}$ to the DGI completion of M is an equivalence.

Proof: Follows directly from Lemma 5.9 above, together with Theorem 4.10 of [8], from which it follows that the DGI completion has an invariance property similar to that proved in Lemma 5.9 above.

MAIN ISOMORPHISM THEOREM

In this section we will prove the following theorem about derived completions.

Theorem 6.1: Suppose that we have a diagram $A \to B \xrightarrow{f} C$ of commutative S-algebras. Suppose further that A, B, and Care all (-1)-connected, and that the homomorphism π_0 (f) is an isomorphism. Suppose further that the natural homomorphism's $\pi_0(A) \to \pi_0(B)$ and $\pi_0(A) \to \pi_0(C)$ are surjections. Then for any connective left A-module spectrum M, the natural homomorphism $MB \to M_C$ is an equivalence of A-module spectra.

The proof of this theorem requires some preliminary technical work on cosimplicial spaces and spectra. Let $X^{\cdot}$ denote any fibrant cosimplicial space (or, more generally, a fibrant cosimplicial spectrum or A-module, where A is a commutative S-algebra), i.e. a space-valued functor from the category Δ whose objects are the totally ordered sets $\underline{k} = \{0,1,...,k\}$ and whose Orphisms are the order-preserving maps of sets. From Proposition 2.6, we have that $\mathrm{Tot}(X')$ is weakly equivalent to $\underset{\underset{\Delta}{\leftarrow}}{\mathrm{holim}} X^{\cdot}$, and similarly that $\mathrm{Tot}^n(X') \cong \underset{\underset{\Delta^{(n)}}{\leftarrow}}{\mathrm{holim}} X^{\cdot}$, where for any non-negative integer, $\Delta^{(n)}$ denotes the full subcategory on the subsets of cardinality $\leq n$. Also, let D_n denote the partially ordered set of non-empty subsets of the set $\underline{n}$, regarded as a category with a unique morphism from S to T whenever $S \subseteq T$, and so that $\mathrm{Hom} D_n(S, T) = \varphi$ whenever $S \not\subseteq T$. For any subset $S \subseteq \underline{n}$ of cardinality $s+1$, let $\pi_n(S) = \underline{s}$. Since S inherits a total ordering from that of $\underline{n}$, there is a unique order-preserving bijective map $\xi_S : S \to \pi_n(S)$. For any inclusion $S \subseteq T$ in D_n, we let $\pi_n(S \subseteq T)$ denote the unique morphism in $\Delta^{(n)}$ which makes the diagram

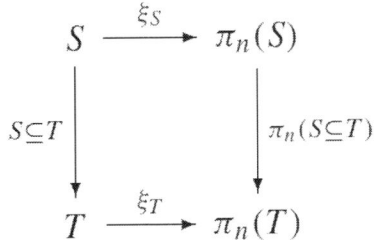

commute. The two definitions make π_n into a functor from D_n to $\Delta^{(n)}$. Our next goal is now to prove that the natural pullback map

$$\pi_n^* : \operatorname*{holim}_{\underleftarrow{\Delta^{(n)}}} X^{\cdot} \to \operatorname*{holim}_{\underleftarrow{\mathcal{D}_n}} X^{\cdot} \circ \pi_n$$

is an equivalence. In order to do this, we recall that by [5], Theorem 9.2, it will suffice to show that for any object $\underline{k} \in \Delta^{(n)}$, the category $\pi_n \downarrow \underline{k}$ has a contractible nerve. Recall that $\pi_n \downarrow \underline{k}$ denotes the category whose objects are pairs (S, θ), where $S \in_{D_n}$ and $\pi_n(S) \to \underline{k}$ is a morphism in $\Delta^{(n)}$, and where a morphism from (S, θ) to (S', θ') in $\pi_n \downarrow \underline{k}$ is a morphism φ from S to S' in D_n making the diagram

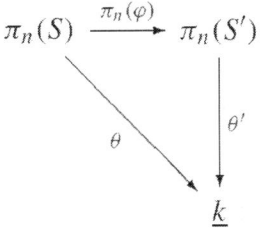

commute. We will construct a category equivalent to $\pi_n \downarrow \underline{k}$ which is readily analyzed. For any finite totally ordered set S, and any positive integer j, a j-fold interval decomposition of S is a partition

$$S = S_0 \coprod S_1 \coprod \cdots \coprod S_j$$

of S, so that whenever $0 \leq i < i' \leq j$, then for any $x \in S_i$ and $y \in S_{i'}$, $x < y$.

Note that some of the sets S_i may be empty. We let P_n^k be the partially ordered set whose objects are pairs $(S, \{S_j\}_{0 \leq j \leq k})$, where S is a non-empty subset of $\underline{n}$, $\{S_j\}_{0 \leq j \leq k}$ is a k-fold interval decomposition of S, and where $(S, \{S_j\}_{0 \leq j \leq k}) \leq (T, \{T_j\}_{0 \leq j \leq k})$ if and only if $S_j \subseteq T_j$ for all $0 \leq j \leq k$.

Proposition 6.2: There is a natural equivalence of categories from the category attached to the poset P_n^k to the category $\pi_n \downarrow \underline{k}$. Consequently, in order to verify that $\alpha : N.P_n^k \to \pi_n \downarrow \underline{k}$ has contractible nerve, it suffices to show that P_n^k does.

Proof

Define $\alpha : P_n^k \to \pi_n \downarrow \underline{k}$ by setting $\alpha(S, \{S_j\}_{0 \leq j \leq k})$ equal to the pair (S, θ), where $\theta : S \to \underline{k}$ is the unique order-preserving map described by $\theta(x) = j$ if and only if $x \in S_j$. It is easy to check that $\pi_n \downarrow \underline{k}$ is a partially ordered set in the sense that for any pair of objects x and y, $\mathrm{Hom} \pi_n \downarrow \underline{k}(x, y)$ is either empty or consists of a single element. One easily sees that the construction α respects the partial orderings, and hence creates a functor. This process is clearly completely reversible (in fact, it is an isomorphism of categories), and the result follows.

Proposition 6.3: $N.P_n^k$ is contractible whenever $k \leq n$.

Proof

We proceed by induction on k. Of course, if $k = 0$, we have that $P_n^0 \cong D_n$, and the latter poset has a maximal element, whence its nerve is contractible. Now, suppose we know the result to be true for all $k \leq K$, and we wish to prove it for K. Let $n \geq K$. The poset P_n^k contains $K + 1$ minimal elements $\{x_0, x_1, \ldots, x_K\}$, where x_j denotes the element whose

underlying set is $\{n\}$, and whose partition places n in the jth interval.
Let $\tilde{P}_n^K \subseteq P_n^K$ denote the partially ordered subset $P_n^K - \{x_0, x_1, ..., x_K\}$. We
also have the obvious embedding $i : P_{n-1}^k \, \tilde{P}_n^K$. We claim that this inclusion induces a homotopy equivalence on nerves. To see this, it suffices
to construct an order-preserving map $f : \tilde{P}_n^K \to P_{n-1}^k$ so that $f \circ i = id$, and
so that $i \circ f(x) \leq x$ for all $x \in \tilde{P}_n^K$. But now we note that the map given
by $(S, \{S_j\}) \to (S - \{n\}, \{S_j - \{n\}\})$ provides the required map of partially ordered sets. For each $0 \leq j \leq K$, we let $Q_j \subseteq \tilde{P}_n^K$ denote the subset
$\{y \in \tilde{P}_n^K \to P_{n-1}^k | y \geq x_i|\}$. As in [15], one now verifies that we have a decomposition

$$N.P_n^K \cong N.\tilde{P}_n^K \cup CN.Q_0 \cup CN.Q_1 \cup \cdots \cup CN.Q_K.$$

We next observe that the inclusion $Q_0 \, \tilde{P}_n^K$ induces an equivalence on
nerves. To see this, we observe that the restriction of the poset map f
constructed above to Q_0 is an isomorphism of partially ordered sets. It
follows that the simplicial set $N.\tilde{P}_n^K \cup CN.Q_0$ is contractible. Now, $N.P_n^K$
is obtained from $N.\tilde{P}_n^K \cup CN.Q_0$ by adjoining cones to the subspaces
$N.Q_1, N.Q_2, ..., N.Q_n$. This means that in order to prove the contractibility of $N.P_n^K$, it will suffice to prove the contractibility of the sets
$N.Q_j$, for $j = 1, 2, ..., n$. We claim that the partially ordered set Q_j is
isomorphic to p_{n-1}^{K-j}. Since we have $K - j \leq n - 1$ whenever $j \geq 1$, we have
that $N.P_{n-1}^{K-j}$ is contractible by the inductive hypothesis, and the result
would follow. To establish the isomorphism $Q_j \cong p_{n-1}^K$, we first note that
$(S, \{S_i\}_{0 \leq i \leq K}) \in Q_j$ if and only if $n \in S_{K-j}$. Therefore, for $(S, \{S_i\}_{0 \leq i \leq K}) \in Q_j$
we have that $S_l = \emptyset$ for $l > K - j$. The isomorphism of partially ordered
sets is now given by $(S, \{S_i\}_{0 \leq i \leq K}) \to (S - \{n\}, \{S_i - \{n\}\}_{0 \leq i \leq K-j})$.

Consider any spectrum-valued functor F on D_n, and for any spectrum X, we let F_X denote the constant functor on D_n with value X. Suppose that we are given a natural transformation $F_X \to F$. We are interested in constructing a useful model for the homotopy fiber of the natural map

$$\operatorname*{holim}_{\mathcal{D}_n} F_X \to \operatorname*{holim}_{\mathcal{D}_n} F.$$

We let I denote the partially ordered set of non-empty subsets of $\underline{1}$, i.e. the category pictured by the diagram

$$\{0\} \longrightarrow \{0, 1\} \longleftarrow \{1\}$$

and let J denote the full subcategory on the two objects $\{0\}$ and $\{0,1\}$. Consider the category I^{n+1}, and its full subcategory J^{n+1}. For any object $\sigma=(S_0, S_1, \ldots, S_n)$ of J^{n+1}, we let $\psi(\sigma) \subseteq \underline{n}$ denote the subset $\{j \in \underline{n} \,|\, S_j| = \{0,1\}\}$. We now define a spectrum-valued functor $\overline{F}$ on I^{n+1} by the formulas

$$\begin{cases} \overline{F}(0, 0, \ldots, 0) = X \\ \overline{F}(\sigma) = F(\psi(\sigma)) \quad \text{for any } \sigma \in J^{n+1} - (0, 0, \ldots, 0) \\ \overline{F}(\xi) = * \quad \text{for any } \xi \in I^{n+1} - J^{n+1}. \end{cases}$$

The behavior on morphisms is evident, when we recall that we are given a natural transformation from the constant functor F_X to F.

Proposition 6.4
There is a natural equivalence (in the homotopy category) ε_n from the homotopy fiber Φ of the natural map to

$$\operatorname*{holim}_{\mathcal{D}_n} F_X \to \operatorname*{holim}_{\mathcal{D}_n} F \text{ to } \operatorname*{holim}_{I^{n+1}} \overline{F}.$$

.

Proof

Φ can clearly be interpreted as the homotopy inverse limit over the category $D_n \times I$ of the functor G defined by

$$\begin{cases} G(S, \{0\}) = X \\ G(S, \{0, 1\}) = F(S) \\ G(S, \{1\}) = *. \end{cases}$$

Define E_n to be the partially ordered set obtained from the product poset $D_n \times I$ by identifying the subset $\{0\} \times D_n$ to a single point ϵ. This means that the object set of E_n is $\{\epsilon\} \cup D_n \times \{0,1\} \cup D_n \times \{1\}$, that $\epsilon \leq (S, \{0,1\})$ for all $S \in D_n$, that ϵ is incomparable with any element of the form $(S, \{1\})$, and that the ordering relationships between any elements of $D_n \times \{\{0,1\}, \{1\}\}$ are identical to those in the original set $D_n \times I$. There is a natural projection $\pi \colon D_n \times I \to E_n$. It is clear from the definition that the functor G factors over π, i.e. that there is a spectrum-valued functor $\overline{G}$ on E_n so that $G = \overline{G} \circ \pi$. On the other hand, we may also define a functor $\rho \colon I^{n+1} \to E_n$ as follows. For any element $v = \{S_0, S1, \ldots, S_n\}$ of $I^{n+1} - J^{n+1}$, we define $\theta(v) \in D_n$ to be the subset $\{j \in \underline{n} \mid S_j \mid = \{1\} \text{ or } \{0,1\}\}$. The functor ρ is now defined by

$$\begin{cases} \rho(\{0\}, \{0\}, \ldots, \{0\}) = \epsilon \\ \rho(S_0, S_1, \ldots S_n) = \psi(S_0, S_1, \ldots, S_n) \times \{0, 1\} \quad \text{for any } (S_0, S_1, \ldots, S_n) \in J^{n+1} - (\emptyset, \emptyset, \ldots, \emptyset) \\ \rho(S_0, S_1, \ldots, S_n) = \theta(S_0, S_1, \ldots, S_n) \times \{1\} \quad \text{for any } (S_0, S_1, \ldots, S_n) \in I^{n+1} - J^{n+1}. \end{cases}$$

Behavior on morphisms is determined since all categories in question are partially ordered sets, and one readily checks that the ordering is respected. One now checks directly that $G \circ \rho$ is identical to the functor $\overline{F}$ defined above. It is also direct to check that for any object $x \in E_n$, the categories $\pi \downarrow x$ and $\rho \downarrow x$ have contractible nerves. The result now follows from [5], Theorem 9.2. $\square$

Now let A be a commutative S-algebra, B a commutative A-algebra spectrum, and let Ψ denote the I-diagram

$$A \to B \leftarrow *$$

of A-module spectra, i.e $\{0\} \to A$, $\{0,1\} \to B$, and $\{1\} \to *$. Then we can define an I^{n+1}-diagram Ψ^n by

$$\Psi^n(S_0, S_1, \ldots, S_n) = \Psi(S_0) \underset{A}{\wedge} \Psi(S_1) \underset{A}{\wedge} \cdots \underset{A}{\wedge} \Psi(S_n).$$

This is a diagram of A-bimodule spectra. We record the following lemma concerning homotopy inverse limits of these I-diagrams.

Lemma 6.5
Let M denote any A-module. Then

$$\underset{I}{\mathrm{holim}}(\Psi(-) \underset{A}{\wedge} M) \cong (\underset{I}{\mathrm{holim}}\ \Psi) \underset{A}{\wedge} M.$$

Proof
We note that by the definitions of homotopy inverse limits, we find that $\underset{I}{\mathrm{holim}}\psi$ is the homotopy fiber of the spectrum map $A \to B$. Consequently, the result reduces to the fact that given any map $f: P \to Q$ of A-module spectra, and any A-module spectrum M, we have an equivalence $\mathrm{Fib}(f) \underset{A}{\wedge} M \cong \mathrm{Fib}(f \underset{A}{\wedge} \mathrm{id}_M)$. But this is clear from the results of [10] and [12]. □

The following result will be the key in proving our main theorem.

Proposition 6.6
Let A and Ψ be as above, and let M be an A-module spectrum. Then we have a natural equivalence

$$\underset{I^{n+1}}{\mathrm{holim}}\ M \underset{A}{\wedge} \Psi^n \cong M \underset{A}{\wedge} \underbrace{\underset{I}{\mathrm{holim}}\ \Psi \underset{A}{\wedge} \cdots \underset{A}{\wedge} \underset{I}{\mathrm{holim}}\ \Psi}_{n+1\ \textit{factors}}.$$

Proof

By induction on n. The result is trivially true for $n=0$. Suppose that it holds for $n=N$, and we wish to prove it for $n=N+1$. We have the string of equivalences

$$\operatorname*{holim}_{(S_0,\ldots,S_{N+1})\in I^{N+2}} (M \wedge_A \Psi^{N+2}(S_0, \ldots, S_{N+1}))$$

$$\cong \operatorname*{holim}_{S_0 \in I} \left[\operatorname*{holim}_{(S_1,\ldots,S_{N+1})\in I^{N+1}} M \wedge_A \Psi(S_0) \wedge_A \Psi^{N+1}(S_1, \ldots, S_{N+2}) \right]$$

$$\cong \operatorname*{holim}_{S_0 \in I} M \wedge_A \Psi(S_0) \wedge_A \left[\underbrace{\operatorname*{holim}_I \Psi \wedge_A \cdots \wedge_A \operatorname*{holim}_I \Psi}_{N+1 \text{ factors}} \right]$$

$$\cong M \wedge_A \left[\operatorname*{holim}_I \Psi \right] \wedge_A \left[\underbrace{\operatorname*{holim}_I \Psi \wedge_A \cdots \wedge_A \operatorname*{holim}_I \Psi}_{N+1 \text{ factors}} \right]$$

$$\cong M \wedge_A \left[\underbrace{\operatorname*{holim}_I \Psi \wedge_A \cdots \wedge_A \operatorname*{holim}_I \Psi}_{N+2 \text{ factors}} \right].$$

The first equivalence is a "Fubini" type theorem for homotopy inverse limits, the second is an application of the inductive hypothesis, and the third is an application of 6.5.

Corollary 6.7

Let $f: A \to B$ be a homomorphism of commutative S-algebras, and let M be a A-module spectrum. Let I denote the homotopy fiber of the homomorphism f. I is an A-module spectrum. Then there is a homotopy fibration sequence of A-module spectra

$$\mathcal{I}_A^{\wedge(k+1)} \underset{A}{\wedge} M \to M \to \text{Tot}^k\, T_A^{\cdot}(M;\, B).$$

Moreover, these sequences fit together in homotopy commutative diagrams

$$
\begin{array}{ccccc}
\mathcal{I}_A^{\wedge(k+2)} \underset{A}{\wedge} M & \longrightarrow & M & \longrightarrow & \text{Tot}^{k+1}(T_A^{\cdot}(M;\, B) \\
\downarrow & & \downarrow{\scriptstyle id_M} & & \downarrow \\
\mathcal{I}_A^{\wedge(k+1)} \underset{A}{\wedge} M & \longrightarrow & M & \longrightarrow & \text{Tot}^k(T_A^{\cdot}(M;\, B)
\end{array}
$$

(6.1)

where the right hand vertical map is the usual projection, and the left hand vertical map is the composite

$$\mathcal{I}_A^{\wedge(k+2)} \cong \mathcal{I} \wedge \mathcal{I}_A^{\wedge(k+1)} \xrightarrow{j \underset{A}{\wedge} id} A \underset{A}{\wedge} \mathcal{I}_A^{\wedge(k+1)} \cong \mathcal{I}_A^{\wedge(k+1)}$$

(6.2)

$j: I \to A$ being the evident inclusion.

This result implies the following technical corollary.

Corollary 6.8

Suppose that we have a homomorphism $f: A \to B$ of (-1)-connected commutative S-algebras, and that $\pi_0(f)$ is surjective. Suppose that

M is a k-connected A-module spectrum. Then each of the spectra $\text{Tot}^n(T_A^{\cdot}(M;B))$ is k-connected, and $M_B = \text{Tot}(T_A^{\cdot}(M;B))$ is at least $k-1-$ connected.

Proof
We prove the first part by induction on n. The result holds for $n=0$ since $\text{Tot}^0(T_A^{\cdot}(M;B)) \cong B \underset{A}{\wedge} M$. The Künneth spectral sequences for smash products show that the connectivity $B \underset{A}{\wedge} M$ is at least k. For higher values of n, we suppose the result known for the value $n-1$. By the commutative diagram (6.1) inCorollary 6.7 above, we have an identification of the homotopy fiber of the map $\text{Tot}^n(T_A^{\cdot}(M;B)) \to \text{Tot}^{n-1}(T_A^{\cdot}(M;B))$ with the homotopy cofiber of the map $I_A^{\wedge(n+1)} \to I_A^{\wedge n}$ in diagram(6.2) above. Since both $I_A^{\wedge(n+1)}$ and $I_A^{\wedge n}$ are (-1)-connected, so is the homotopy fiber of the map $\text{Tot}^n(T_A^{\cdot}(M;B)) \to \text{Tot}^{n-1}(T_A^{\cdot}(M;B))$. It now readily follows that $\text{Tot}^n(T_A^{\cdot}(M;B))$ is (-1)-connected. The result for $\text{Tot}(T_A^{\cdot}(M;B))$ now follows from the $\lim^1$-exact sequence for homotopy fibrations (see [5], Ch X, Section 3).

We will also need the following technical lemma.

Lemma 6.9

Let $A \xrightarrow{f} B \to C$ be a diagram of commutative S-algebras, and let M denote an A-module spectrum. Suppose further that there is a homomorphism of commutative A-algebra spectra $s: C \to B$ such that $s \circ f = id_B$.

Then the natural map $M_B^{\wedge} \to M_C^{\wedge}$ is an equivalence of spectra.

Proof

Follows directly from Proposition 2.15, applied to the triples $s = B \wedge_A -$

and $c \wedge_A -$.

Proof of Theorem 6.1

We consider first the case when the diagram of S-algebra homomorphisms is $A \to A \to C$, i.e. $A = B$, and where the homotopy fiber I of the map $f : A = B \to C$ is 0-connected, i.e. when $\pi_0(f)$ is an isomorphism and $\pi_1(f)$ is surjective. It follows from Proposition 6.4 and Proposition

6.6 that the homotopy fiber F of the map $\underset{\overset{\longleftarrow}{\Delta^R}}{M \to \text{holim}\, T_A(M;B) \cong \text{Tot}^n T_A(M;B)}$ is equiv-

alent to $\underbrace{I \wedge_A \cdots \wedge_A I}_{n+1 \text{ factors}}$. Since I is 0-connected, it follows by an easy applica-

tion of Corollary 2.5 that F is n-connected. This gives the result in this case, since when one passes to the homotopy limit over n the connectivity of the map goes to infinity, and since $M \wedge_A \cong M$.

We next consider the case where A is not necessarily equal to B, but where we have that the homotopy fiber I of the map $f : B \to C$ is 0-connected. We construct the bicosimplicial spectrum $C^{\cdot\cdot}$ given by $C^{pq} = T_A^p(T_A^p(M;B);C)$, with the obvious coface and codegeneracy maps. Lemma 2.14 asserts that the natural map

$$T_A(M;C) \to C^{\cdot\cdot}$$

(induced by $\eta : M \to T_A(M;B)$) induces an equivalence on total spectra, where $T_A(M;C)$ is regarded as a bicosimplicial spectrum constant in the q-direction. We also have the natural map $\eta : T_A(M;B) \to C^{\cdot\cdot}$ obtained by regarding $\eta : M \to T_A(M;B)$ as a bicosimplicial spectrum constant in the p-direction. Since $\text{Tot}(C^{\cdot\cdot}) \cong M_C^\wedge$, and this equivalence is compatible with the standard map $M_B^\wedge \to M_C^\wedge$, it suffices to show that η induces a weak equivalence on total spectra. For this it suffices to show that each of the maps

$$T_A^q(M; B) \to \mathrm{Tot}(T_A(T_A^q(M; B)); C)$$

is a weak equivalence of spectra. Since each of the A-module spectra

$$T_A^q(M; B) = \underbrace{B \wedge_A \cdots \wedge_A B \wedge_A M}_{q+1 \text{ factors}}$$

is equipped with a B-module structure restricting to its given A-module structure, it will now suffice to prove that the natural map $M_B^\wedge \to M_C^\wedge$ is an equivalence for any B-module spectrum M. For a B-module spectrum M, we may construct the cosimplicial spectrum $T_B(M;C)$. From the hypothesis on f, and by the previous case, we find that the connectivity of the natural maps $M \to (\mathrm{Tot}^n)(T_B(M;C))$ goes to infinity with n. We will now apply the functors $T_A(-;B)$ and $T_A(-;C)$ levelwise to the cosimplicial spectrum $T_B(M;C)$ to obtain bicosimplicial spectra, with a natural bicosimplicial map from the first to the second induced by the A-algebra spectrum map $B \to C$. We obtain a commutative diagram

$$
\begin{array}{ccc}
M_B^\wedge = \mathop{\mathrm{Tot}}_{p \in \Delta} T_A^p(M; B) & \longrightarrow & M_C^\wedge = \mathop{\mathrm{Tot}}_{p \in \Delta} T_A^p(M; C) \\
\downarrow & & \downarrow \\
\mathop{\mathrm{Tot}}_{p \in \Delta} \mathop{\mathrm{Tot}}_{q \in \Delta} T_A^p(T_B^q(M; C); B) & \longrightarrow & \mathop{\mathrm{Tot}}_{p \in \Delta} \mathop{\mathrm{Tot}}_{q \in \Delta} T_A^p(T_B^q(M; C); C).
\end{array}
$$

$$(6.3)$$

We wish to prove that the upper horizontal arrow is an equivalence of spectra. We will do this by showing that the two vertical arrows and the lower horizontal arrow are equivalences of spectra. We first note that

by the increasing connectivity (with n) of the maps $M \to \mathrm{Tot}^n_{q \in \Delta}(T^q_B(M;C))$ and Corollary 6.8, we find that the natural maps

$$\mathrm{Tot}_{p \in \Delta} T^p_A(M;B) \to \varprojlim_n \mathrm{Tot}_{p \in \Delta} T^p_A(\mathrm{Tot}^n_{q \in \Delta} T^q_B(M;C);B)$$

and

$$\mathrm{Tot}_{p \in \Delta} T^p_A(M;C) \to \varprojlim_n \mathrm{Tot}_{p \in \Delta} T^p_A(\mathrm{Tot}_{q \in \Delta^n} T^q_B(M;C);C).$$

are equivalences of spectra. By Proposition 2.9, we have that

$$T^p_A(\mathrm{Tot}^n_{q \in \Delta} T^q_B(M;C);B) \cong \mathrm{Tot}^n_{q \in \Delta} T^p_A(T^q_B(M;C);B)$$

and

$$T^p_A(\mathrm{Tot}^n_{q \in \Delta} T^q_B(M;C);C) \cong \mathrm{Tot}^n_{q \in \Delta} T^p_A(T^q_B(M;C);C).$$

Therefore, the maps

$$\mathrm{Tot}_{p \in \Delta} T^p_A(M;B) \to \varprojlim_n \mathrm{Tot}_{p \in \Delta} \mathrm{Tot}^n_{q \in \Delta} T^p_A(T^q_B(M;C);B) \cong \mathrm{Tot}_{p \in \Delta} \mathrm{Tot}_{q \in \Delta} T^p_A(T^q_B(M;C);B)$$

and

$$\mathrm{Tot}_{p \in \Delta} T^p_A(M;C) \to \varprojlim_n \mathrm{Tot}_{p \in \Delta} \mathrm{Tot}^n_{q \in \Delta} T^p_A(T^q_B(M;C);C) \cong \mathrm{Tot}_{p \in \Delta} \mathrm{Tot}_{q \in \Delta} T^p_A(T^q_B(M;C);C)$$

are also weak equivalences of spectra. These are the vertical arrows in Diagram (6.3). We must also check that the lower horizontal arrow in Diagram (6.3) is an equivalence of spectra. This arrow is of course induced by a map of bicosimplicial spaces, so we may verify it by proving that it is a levelwise equivalence. Specifically, if we can show that for each q, the map

$$\operatorname*{Tot}_{p \in \Delta} T_A^p(T_B^q(M;C);B) \rightarrow \operatorname*{Tot}_{p \in \Delta} T_A^p(T_B^q(M;C);C)$$

is an equivalence of spectra, then the lower horizontal map in Diagram

(6.3) will be an equivalence. But, $T_B^q(M;C)$ admits a C-module structure (and therefore also a B-module structure), so by Proposition 3.2,(6), we have

$$\operatorname*{Tot}_{p \in \Delta} T_A^p(T_B^q(M;C);B) \cong \operatorname*{Tot}_{p \in \Delta} T_A^p(T_B^q(M;C);C)$$

which gives the result, i.e. that the statement of the theorem holds in the case when I is 0-connected.

The general case, i.e. without the assumption that $\pi_1 B \rightarrow \pi_1 C$ is surjective, can now be handled as follows. Let $\{f\alpha\}_{\alpha \in a}$ be a generating set for $\pi_1 C$. Let $Z = V_{\alpha \in A} S_\alpha^1$ denote the suspension spectrum of a bouquet of circles parametrized by A, and let $\theta: Z \rightarrow C$ denote the map whose restriction to S_α^1 is the map f_α. Let B (Z) denote the free B-module on the spectrum Z. Then we obtain a map of commutative A-algebra spectra $g : \operatorname{Sym}_B(B(Z)) \rightarrow C$. One readily checks that $\pi_0(g)$ is an isomorphism and that $\pi_1(g)$ is surjective, so we may conclude that $\operatorname{Mym}_B(B(Z))^{\rightarrow M_C}$ is an equivalence of spectra. On the other hand, the inclusion $B \rightarrow \operatorname{Sym}_B(B(Z))$ admits a section, and so the natural map $M_B \rightarrow \operatorname{Msym}_{B(B(Z))}$ is also an equivalence by6.9. This concludes the proof of Theorem 6.1.

We have been discussing the invariance of completion along an S-algebra homomorphism f: A→B under changes in the target S-algebra B. We will also need a similar result which will demonstrate invariance under changes in A under suitable circumstances. For a homomorphism of commutative S-algebrasf: A→B, and a B-module spectrum M, we will denote by $\rho^A M$ the result of regarding M as an A-module spectrum via the S-algebra homomorphism f.

Theorem 6.10

Let $A \xrightarrow{f} B \to C$ be a diagram of (-1)-connected commutative S-algebras, and let M be a B-module spectrum. Suppose that the homomorphism $\pi_0 f : \pi_0 A \to \pi_0 B$ is an isomorphism. Then the natural homomorphism $(\rho^A M)_C^{\wedge} \to M_C^{\wedge}$ is a weak equivalence of spectra.

Proof

Let T_A and T_B denote the triples $M \mapsto C \wedge_A M$ and $M \mapsto C \wedge_B M$, respectively. We consider the bicosimplicial A-module spectrum

$$(p, q) \mapsto T_B^p T_A^q (M)$$

as usual, and we have by Lemma 2.14 that its total space is equivalent to $M_C^{\wedge}$. In order to check the equivalence we are interested in, it will be sufficient to prove that for each q, the total spectrum of the cosimplicial spectrum

$$p \mapsto T_B^p T_A^q (M)$$

is weakly equivalent to $T_A^q(M)$ under the natural map $\eta_B(T_A^q(M))$. As in the proof of Theorem 6.1, we let I denote the homotopy fiber of the homomorphism $B \to C$. I is a B-module. Then Corollary 6.7 shows that it will suffice to prove that the inverse system of spectra

$$\cdots \to I_B^{\wedge k+1} \wedge_B T_A^p(M) \to I_B^{\wedge k} \wedge_B T_A^p(M) \to I_B^{\wedge k-1} \wedge_B T_A^p(M) \to \cdots$$

has contractible homotopy inverse limit. It is clear that to verify this, it will suffice to show that for any B-module spectrum N, the inverse system of spectra

$$\cdots \to \mathcal{I}_B^{\wedge^{k+1}} \underset{B}{\wedge} C \underset{A}{\wedge} N \to \mathcal{I}_B^{\wedge^k} \underset{B}{\wedge} C \underset{A}{\wedge} N \to \mathcal{I}_B^{\wedge^{k-1}} \underset{B}{\wedge} C \underset{A}{\wedge} N \to \cdots$$

has contractible homotopy inverse limit. In order to prove this, we observe that there is a standard equivalence

$$C \underset{A}{\wedge} N \cong C \underset{B}{\wedge} B \underset{A}{\wedge} B \underset{B}{\wedge} M$$

of B-module spectra. The idea will be to replace $B \underset{A}{\wedge} B$ by its cosimplicial approximation $T^{\cdot}_{B \wedge B}(B \underset{A}{\wedge} B; B)$, where B is a commutative $B \underset{A}{\wedge} B$-algebra via the multiplication map $B \underset{A}{\wedge} B \to B$. Since we are assuming that $\pi_0 f$ is an isomorphism, it follows that the map $\pi_0 B \underset{A}{\wedge} B \to \pi_0 B$ is an isomorphism, and therefore by Theorem 6.1 that the natural map η gives an equivalence $B \underset{A}{\wedge} B \to (B \underset{A}{\wedge} B)^{\wedge}_B$. Also, $\pi_1 B \underset{A}{\wedge} B \to \pi_1 B$ is surjective since the multiplication map has a section. It follows that the natural map $B \underset{A}{\wedge} B \to \mathrm{Tot}^j(T^{\cdot}_{B \wedge B}(B \underset{A}{\wedge} B; B))$ is at least $(i-1)$-connected. Letting $T^i = \mathrm{Tot}^j(T^{\cdot}_{B \wedge B}(B \underset{A}{\wedge} B; B))$, it now follows from the Künneth spectral sequence that the natural map $C \underset{A}{\wedge} N \to C \underset{B}{\wedge} T^i \underset{B}{\wedge} N$ is $(i-1)$-connected. Consequently, we have an equivalence

$$C \underset{A}{\wedge} N \cong \mathrm{Tot}(k \mapsto C \underset{B}{\wedge} T^k_{B \underset{A}{\wedge} B}(B \underset{A}{\wedge} B; B) \underset{B}{\wedge} N).$$

Therefore, it will now suffice to prove that the homotopy inverse limit of the system

$$\cdots \to \mathcal{I}_B^{\wedge^{k+1}} \underset{B}{\wedge} C \underset{B}{\wedge} T^k_{B \underset{A}{\wedge} B}(B \underset{A}{\wedge} B; B) \underset{B}{\wedge} N \to \mathcal{I}_B^{\wedge^k} \underset{B}{\wedge} C \underset{B}{\wedge} T^k_{B \underset{A}{\wedge} B}(B \underset{A}{\wedge} B; B) \underset{B}{\wedge} N \to$$

$$\to \mathcal{I}_B^{\wedge^{k-1}} \underset{B}{\wedge} C \underset{B}{\wedge} T^k_{B\underset{A}{\wedge}B}(B\underset{A}{\wedge} B;\, B) \underset{B}{\wedge} N \to \cdots$$

is contractible. Now, $T^k_{B\underset{A}{\wedge}B}(B\underset{A}{\wedge}B;B)$ is a $(k+1)$-fold smash product over

$B\underset{A}{\wedge}B$ of copies of B. By including B into the first tensor factor, we find that it can be viewed as a B-algebra, for which we write $\wedge$. $\wedge$ can be viewed as a B–B-bimodule, with both right and left actions coming from the (left) action of B on $\wedge$. We can now resolve $\wedge$ by free B-modules (each of which is regarded as a bimodule via the left action, and which can be taken to be free on a wedge of spheres of the same dimension), and it is now easy to see that it will suffice to show that the inverse system

$$\cdots \to \mathcal{I}_B^{\wedge^{k+1}} \underset{B}{\wedge} C \underset{B}{\wedge} F \underset{B}{\wedge} N \to \mathcal{I}_B^{\wedge^{k}} \underset{B}{\wedge} C \underset{B}{\wedge} F \underset{B}{\wedge} N \to \mathcal{I}_B^{\wedge^{k-1}} \underset{B}{\wedge} C \underset{B}{\wedge} F \underset{B}{\wedge} N \to \cdots$$

has contractible homotopy inverse limit, where F is any free B-module on a bouquet of spheres of a fixed dimension k, regarded as a B–B-bimodule, with both actions agreeing with the left action on F. But to check this, it will suffice to show that the map

$$\mathcal{I}_B^{\wedge^{k}} \underset{B}{\wedge} C \underset{B}{\wedge} F \underset{B}{\wedge} N \to \mathcal{I}_B^{\wedge^{k-1}} \underset{B}{\wedge} C \underset{B}{\wedge} F \underset{B}{\wedge} N$$

is null homotopic as a map of spectra. But this question clearly can be reduced to the case where $F=B$, so we will have to show that the map

$$\mathcal{I}_B^{\wedge^{k}} \underset{B}{\wedge} C \underset{B}{\wedge} N \to \mathcal{I}_B^{\wedge^{k-1}} \underset{B}{\wedge} C \underset{B}{\wedge} N$$

is null homotopic. This can clearly be reduced to the case where $k=1$, so we need to check that the map

$$\mathcal{I} \underset{B}{\wedge} C \underset{B}{\wedge} N \to B \underset{B}{\wedge} C \underset{B}{\wedge} N \cong C \underset{B}{\wedge} N$$

is null. But we have the cofiber sequence

$$\mathcal{I} \wedge_B C \wedge_B N \to C \wedge_B N \to C \wedge_B C \wedge_B N$$

and the second map is the inclusion of a wedge summand, since we have the retraction

$$\mu \wedge_B id_N : C \wedge_B C \wedge_B N \to C \wedge_B N$$

where $\mu : C \wedge_B C \to C$ is the multiplication map. This gives the result. $\square$

ALGEBRAIC TO GEOMETRIC SPECTRAL SEQUENCE

In this section, we will provide a computational device which will permit the computation of homotopy groups of derived completions of module spectra over commutative S-algebras in terms of derived completions over actual rings. Here is the statement of the main theorem of this section.

Theorem 7.1.
Let $A \to B$ be a map of commutative (-1)-connected S-algebras, with $\pi_0 A \to \pi_0 B$ surjective, and let M be a connective A-module spectrum. $\pi_0 A$ is a commutative ring, $\pi_0 B$ is a commutative $\pi_0 A$-algebra, and for each i, $\pi_i M$ is a $\pi_0 A$-module. We may therefore construct the derived completion spectrum $(\pi_i M)\pi_0 B$ for each i. There is a second quadrant spectral sequence with $E_1^{pq} = \pi_{q+2p}((\pi_{-p} M^{\wedge}_{\pi_0 B}))$, converging to $\pi_{p+q}(M_B)$.

We will require some preliminary technical work on Postnikov decompositions of S-algebras and module spectra before we can present the proof of this result. We recall from [17] and [18] that the category Alg_s admits the structure of a Quillen model category, in which the weak equivalences are the homomorphisms of S-algebras inducing isomorphisms on homotopy groups. There is also a "free commutative S-algebra" on a spectrum X, which we denote by $\mathrm{Sym}(X)$. This construction satisfies the adjointness relationship

$$\mathrm{Hom}_{\mathrm{Alg}_S}(\mathrm{Sym}(X), A) \cong \mathrm{Hom}_{\mathrm{Mod}_S}(X, A).$$

Sym is a triple on the category of spectra, and commutative S-algebras are exactly the algebras over the triple Sym. For any commutative S-algebra A, we may as in Definition 2.16 construct its simplicial resolution Sym.(A), relative to the triple Sym. Moreover, Proposition 2.17 shows that the natural map Sym.(A)→A is a weak equivalence. In [18], it is shown that Sym.(A) is a cofibrant object in the model structure on the category Alg_S.

Given two commutative S-algebras A and B, the set $\mathrm{Hom}_{\mathrm{Alg}_S}(A,B)$ is equipped with the structure of a space, as a subspace of the zeroth space of the spectrum $\mathrm{Hom}_{\mathrm{Mod}_S}(A,B)$. This construction is not homotopy invariant, but the result of replacing A by any weakly equivalent cofibrant object does yield a homotopy invariant notion. Of course, we may take the cofibrant replacement to be Sym.(A). We now need a result about the space of S-algebra maps to Eilenberg–MacLane spectra.

Proposition 7.2
Let A be a (−1)-connected commutative S-algebra. Let B denote any commutative ring, and H (B) the corresponding Eilenberg–MacLane spectrum. The canonical map from the space

$$\mathrm{Hom}_{\mathrm{Alg}_S}(\mathrm{Sym}.(A), H(B)) \rightarrow \mathrm{Hom}(\pi_0(A), B)$$ is an equivalence, i.e.

$\mathrm{Hom}_{\mathrm{Alg}_S}(\mathrm{Sym}.(A), H(B))$ is a space whose components are in bijective correspondence with the ring homomorphisms from $\pi_0(A)$ to B, and so that each component is contractible.

Proof
The adjunction $\mathrm{Hom}_{\mathrm{Alg}_S}(\mathrm{Sym}(X), A) \cong \mathrm{Hom}_{\mathrm{Mod}_S}(X, A)$ shows that the space

$\text{Hom}_{\text{Alg}_S}(\text{Sym}.(A), \mathbb{H}(B))$

is equivalent to the total space of the cosimplicial space which in level k is the space of spectrum maps from $\text{Sym}^k(A)$ to $H(B)$. Since the $k+1$-skeleton $Sk^{(k+1)}\text{Sym}.(A)$ is obtained from the k-skeleton by attaching cells in dimensions $k+1$ and higher, it follows that the restriction map

$$\text{Hom}_{\text{Alg}_S}(\text{Sym}.(A), \mathbb{H}(B)) \rightarrow \text{Hom}_{\text{Alg}_S}(Sk^1\text{Sym}.(A), \mathbb{H}(B)) \overset{\text{def}^n}{=} T^1$$

is an equivalence. Now, T^1 is clearly the homotopy equalizer of the two maps

$$\varphi, \psi : \text{Hom}_{\text{Mod}_S}(A, \mathbb{H}(B)) \rightarrow \text{Hom}_{\text{Mod}_S}(\text{Sym}(A), \mathbb{H}(B))$$

defined by

$$\begin{cases} \Phi(f) = f \circ \alpha_{\pi_0 A} \text{ where } \alpha_{\pi_0 A} : \text{Sym}(\pi_0 A) \rightarrow \pi_0 A \text{ is the structure map for the ring } \pi_0 A. \\ \Psi(f) = \alpha_B \circ \text{Sym}(f) \text{ where } \alpha_B : \text{Sym}(B) \rightarrow B \text{ is the structure map for the ring } B. \end{cases}$$

The space $\text{Hom}_{\text{Mod}_s}(A, \mathbb{H}(B))$ is clearly equivalent to the discrete space $\text{Hom}_{\text{Ab}}(\pi_0(A), (B)$. Similarly, $\text{Hom}_{\text{Mod}_s}(\text{Sym}(A), \mathbb{H}(B))$ is equivalent to the discrete space

$$\text{Hom}_{\text{Ab}}(\pi_0(\text{Sym}(A)), B) = \text{Hom}_{\text{Ab}}(\text{Sym}(\pi_0(A), B)).$$

The homotopy equalizer is now equivalent to the equalizer of the pair of set maps

$$\Phi, \Psi : \text{Hom}_{\text{Ab}}(\pi_0(A), B) \rightarrow \text{Hom}_{\text{Ab}}(\text{Sym}(\pi_0(A), B))$$

defined by

$$\begin{cases} \Phi(f) = f \circ \alpha_{\pi_0 A} \text{ where } \alpha_{\pi_0 A} : \text{Sym}(\pi_0 A) \rightarrow \pi_0 A \text{ is the structure map for the ring } \pi_0 A. \\ \Psi(f) = \alpha_B \circ \text{Sym}(f) \text{ where } \alpha_B : \text{Sym}(B) \rightarrow B \text{ is the structure map for the ring } B. \end{cases}$$

This set is the set of all abelian group homomorphisms f: $\pi_0 A \to B$ making the diagram

$$
\begin{array}{ccc}
\mathrm{Sym}(\pi_0 A) & \xrightarrow{\mathrm{Sym}(f)} & \mathrm{Sym}(B) \\
\downarrow{\scriptstyle \alpha_{\pi_0 A}} & & \downarrow{\scriptstyle \alpha_B} \\
\pi_0 A & \xrightarrow{\;f\;} & B
\end{array}
$$

commute. This is clearly the set of ring homomorphisms from $\pi_0 A$ to B.

Corollary 7.3

There is a canonical homotopy equivalence class of homomorphisms of commutative S-algebras π_A: $A \to H(\pi_0 A)$ which induces the identity on π_0.

We also have the following results on A-module spectra. They follow immediately from Proposition 3.9 of [8].

Proposition 7.4

Let A be a (−1)-connected commutative S-algebra, and let M be a cofibrant (k−1)-connected A-module spectrum. Let N be any module over the ring $\pi_0 A$, and H (N,k) the k-dimensional Eilenberg–MacLane spectrum for N, regarded as an A-module spectrum via the homomorphism $A \to \pi_0 A$. The zeroth space of the function spectrum

$\mathrm{Hom}_A (M, \mathbb{H}(N,k))$ is homotopy equivalent to the set $\mathrm{Hom}_{\pi_0 A}(\pi_k M, N)$ via the natural map which assigns to any map its induced map on π_k.

Corollary 7.5

Let A be a (−1)-connected commutative S-algebra. Let M be a module spectrum over A, and suppose that the underlying spectrum of M is an Eilenberg–MacLane spectrum, say of dimension n. $\pi_n M$ is a module over the ring $\pi_0 A$. We let Π_M denote the A-module spectrum obtained from $\pi_n M$ by pullback along the ring homomorphism π_A. Then, M is equivalent as an A-module spectrum to Π_M.

These results allow us to construct a Postnikov tower in the category of connective modules over A.

Corollary 7.6

Let A be a (-1)-connected commutative S-algebra, and let M be a k-connected A-module spectrum, for some integer k. Then there is a tower of fibrations of A-module spectra

$$\cdots M[s] \to M[s] \to \cdots \to M[k+1] \to M[k]$$

together with A-module spectrum maps\ $M \xrightarrow{f_s} M[S]$ making all the tri-angles

$$M \xrightarrow{f_s} M[s]$$

$$f_{s-1} \searrow \qquad \downarrow$$

$$M[s-1]$$

commute, with the following properties.

- f_s induces an isomorphism on π_i for $i \leq s$.
- $\pi_i M[s] = 0$ for $i > s$.
- The homotopy fiber of the map $M[s] \to M[s-1]$ is an Eilenberg–MacLane spectrum in dimension s.

- Using the maps $\{f_s\}_{s'}$, $\overset{\text{holim} M[s]}{\underset{s}{\longleftarrow}}$ is naturally equivalent to M

Moreover, the tower is natural in the homotopy category.

Proof

Immediate from Corollary 7.5 above.

We will use the Postnikov tower to construct the spectral sequence. A preliminary observation is the following.

Proposition 7.7

Let

$$\cdots \to M_s \to \cdots \to M_1 \to M_0$$

be any diagram of module spectra over a commutative S-algebra A, and let P denote any cofibrant A-module spectrum. Suppose further that the connectivity of the M_s goes to infinity with s, and that P is connective, i.e. is k-connected for some k. Then the natural map

$$P \underset{A}{\wedge} (\operatorname*{holim}_{s} M_s) \longrightarrow \operatorname*{holim}_{s} (P \underset{A}{\wedge} M_s)$$

is an equivalence of spectra.

Proof

Follows directly from the fact that $P \underset{A}{\wedge}$ preserves connectivity when P is cofibrant.

Corollary 7.8

Suppose that A is a (-1)-connected commutative S-algebra, and that we have a homomorphism $f \colon A \to B$ of commutative S-algebras. Suppose further that B is (-1)-connected and cofibrant. For any diagram

$$\cdots \to M_s \to \cdots \to M_1 \to M_0$$

of connective A-module spectra, for which the connectivity of M_s goes to infinity with s, the natural map

$$(\operatorname*{holim}_{s} M_s)^{\wedge}_B \to \operatorname*{holim}_{s} (M_s)^{\wedge}_B$$

is a weak equivalence of spectra. In particular, if $\{M[s]\}_s$ is the Post-

nikov tower for an A-module spectrum M, then $M_B^\wedge \cong \underset{s}{\text{holim}} M[s]_B^\wedge$. Further,

the homotopy fiber of the map $M[s]_B^\wedge \to M[s-1]_B^\wedge$ is equivalent to $(F_s)_B^\wedge$, where F_s is the homotopy fiber of the map $M[s] \to M[s-1]$.

The spectral sequence in question is simply the spectral sequence on homotopy groups attached to an inverse system of spectra. This is a standard result which is discussed in [5], Chapter IX.

Proposition 7.9
Let $\cdots X_s \to X_{s-1} \to \cdots \to X_1 \to X_0$ be an inverse system of connective spectra, and let F_s denote the homotopy fiber of the map $X_s \to X_{s-1}$. Suppose further that the connectivity of the spaces goes to infinity with s. Then

there exists a left half plane spectral sequence with E_1^{pq}-term $\pi_{p+q} F_{-p}$,

converging to $\underset{s}{\underset{\leftarrow}{\pi_{p+q}(\text{holim} X_s)}}$.

Proof of Theorem 7.1
We apply Proposition 7.9 and Corollary 7.8 to the inverse system

$$\cdots \to M[s]_B^\wedge \to M[s-1]_B^\wedge \to \cdots \to M[k+1]_B^\wedge \to M[k]_B^\wedge$$

where k is the connectivity of M. The homotopy fiber of the map

$M[s]_B^\wedge \to M[s-1]_B^\wedge$ is $(F_s)_B^\wedge$, where as above F_s denotes the homotopy fiber of the map $M[s] \to M[s-1]$. Since $\{M[s]\}_s$ is the Postnikov tower for M, F_s is an s-dimensional Eilenberg–MacLane spectrum with $\pi_s(F_s) \cong \pi_s(M)$. Consequently, by Corollary 7.5, it is equivalent to a module over the ring $\pi_0(A)$, and by Theorem 6.10, it follows that the derived completion of F_s at the S-algebra homomorphism $A \to B$ is equivalent to its derived completion at the ring homomorphism $\pi_0 A \to \pi_0 B$. The theorem now follows.

We can now use this spectral sequence to obtain two useful corollaries. We first have an extension of Proposition 3.2, item 6.

Corollary 7.10

Let $B \to C$ be a homomorphism of (-1)-connected commutative S-algebras, and suppose $\pi_0 B \to \pi_0 C$ is surjective. Let M be a B-module spectrum, and suppose that for each i, the $\pi_0 B$-module structure on $\pi_1 M$ extends to a $\pi_0 C$-action. Then the natural map $\eta : M \to M_C^{\wedge}$ is an equivalence.

Proof

Follows directly from Proposition 3.2, item 6, and Theorem 7.1.

We also have the following corollary, which will be useful in the K-theoretic applications.

Corollary 7.11

Let

$$A \to B \to C$$

be a diagram of (-1)-connected commutative S-algebras, and suppose that $\pi_0 A \to \pi_0 B$ and $\pi_0 B \to \pi_0 C$ are surjective. For any B-module M, let $\rho_A(M)$ denote M regarded as an A-module by restriction of scalars.

Then the natural map $(\rho_A(M))_C^{\wedge} \to M_C^{\wedge}$ is a weak equivalence.

Proof

It follows from Theorem 7.1 that it suffices to consider the case where A, B, and C are all obtained by applying the Eilenberg–MacLane construction to diagram of actual commutative rings, and that M is obtained as an actual module over the ring B. Exactly as in the proof of Theorem 6.10, it suffices to prove that the natural map

$$C \wedge_A M \longrightarrow (C \wedge_A M)_C^{\wedge}$$

is an equivalence. Here the completion is taking place along the homomorphism $B \to C$, and the B-module structure on $C \wedge_A M$ is obtained
from the original B-action on M, not via restriction of scalars from the
extended left C-action. That is, we have $b \cdot (c \wedge m) = c \wedge b \cdot m$. Now, the
homotopy groups of $C \wedge_A M$ are the groups $\mathrm{Tor}^A_*(M, C)$, via the (collapsing) Künneth spectral sequence. We must only show that the B-action
on these Tor-groups obtained from the given B-action on M extends over
C. But from the surjectivity of the homomorphism $A \to B$, the B-action
can be computed as the action of A/J, where J is the kernel of the ring
homomorphisms $A \to B$, and hence is determined by the A-action. It is
clear from the definition that the A-action extends over C, and hence
that it vanishes on the I, the kernel of the ring homomorphism $A \to C$.
But, $A/I \cong B/I'$, where I' is the kernel of $B \to C$, which gives the result.

EXAMPLES

In this section, we will discuss a number of examples of this construction.

Example 8.1

We consider the case where the S-algebras A and B are the Eilenberg–
MacLane spectra for the rings Z and F_p, and where the homomorphism $f : A \to B$ is induced by reduction mod p. For a finitely generated
A-module M, Theorem 4.4 tells us that the derived completion of M
along f is simply the Eilenberg–MacLane spectrum for the completed
module $M^\wedge_p$.

Example 8.2

Let $A = S^0$, the sphere spectrum, and as in the previous example let B
be the Eilenberg–MacLane spectrum for the ring F_p. Let f be the composition of the natural map the the Eilenberg–MacLane spectrum for
Z with mod-p reduction. Any spectrum is in a natural way a module
over the commutative S-algebra S^0, and its completion along this map

is the usual p-adic completion. The tower of fibrations mentioned in Remark 3.1 is in this case identical to the tower in the Adams spectral sequence.

Example 8.3

Let A be the S-algebra ku, i.e. connective complex K-theory, and let B be the Eilenberg–MacLane spectrum for the ring $\mathbb{Z}$. We view ku as the spectrum associated with the topological symmetric monoidal category of finite dimensional complex vector spaces, and we define f to be the homomorphism of S-algebras induced by the functor which sends every complex vector space to its dimension. Since $\pi_0 f$ is an isomorphism, Theorem 7.1 shows that in this case, $ku_B^\wedge$ is equivalent to ku itself. The tower of fibrations mentioned in Remark 3.1 is in this case the accelerated Postnikov tower for ku, with the ith spectrum in the tower of fibrations equivalent to the Postnikov cover $ku[2i, \ldots, +\infty)$

Example 8.4

Let A denote the Eilenberg–MacLane spectrum for the representation ring $R[\mathbb{Z}_p]$, where $\mathbb{Z}_p$ denotes the group of p-adic integers. For a profinite group G, the representation ring $R[G]$ is defined to be the direct limit of $R[G/N]$, where N ranges over all the finite index normal subgroups of the group G. As a ring, one readily sees that $R[\mathbb{Z}_p] \cong \mathbb{Z}[\mathbb{Z}/p^\infty \mathbb{Z}]$, where $\mathbb{Z}[-]$ denotes integral group ring, and where $\mathbb{Z}/p^\infty \mathbb{Z}$ denotes the union $U_n \mathbb{Z}/p^n \mathbb{Z}$. We let $C = F_p$, and define $f : A \to C$ to be the composite $R[\mathbb{Z}_p] \to R[\{e\}] = \mathbb{Z} \to F_p$. We will let the module M be the ring A itself. We let S^1 denote the simplicial group of singular chains on the circle group, with multiplication and inverse induced by those on the topological group. There is of course a natural homomorphism of simplicial groups $\dfrac{\mathbb{Z}/P^\infty \mathbb{Z} \to S^1}{HF}$, induced by the identification of $\mathbb{Z}/p^\infty \mathbb{Z}$ with the group of p-power torsion points on S^1, and where $\mathbb{Z}/p^\infty \mathbb{Z}$ is regarded as a discrete simplicial group. We define B to be the simplicial group ring $\mathbb{Z}[S^1]$, which is a simplicial algebra $\mathbb{Z}[\mathbb{Z}/p^\infty\mathbb{Z}]$. Such a simplicial algebra may be regarded as a commutative S-algebra, in fact a commutative algebra spectrum over the S-algebra $\mathbb{H}(A)$. We let $\rho(B)$ denote B regarded as an A-module spectrum via restriction of scalars along the inclusion $\mathbb{H}(A) \to B$. We then have a composite

$$\mathbb{H}(A)^\wedge_C \to \rho(B)^\wedge_C \to B^\wedge_C. \tag{8.4}$$

Note that the right hand spectrum is obtained by completion over the ground S-algebra B, and the two leftmost spectra are completions over the ground S-algebra $\mathbb{H}(A)$. The right hand map is induced by the natural morphism of cosimplicial spectra $T_{\mathbb{H}(A)}(\rho(B);C) \to T_B(B;C)$, so the composite is induced by the natural morphism $T_{\mathbb{H}(A)}(A;C) \to T_B(B;C)$. One can now check from the definitions that

$$\pi_s(T^k_A(A;C)) \cong \mathrm{MultiTor}^A_s(\underbrace{\mathbb{F}_p, \mathbb{F}_p, \ldots, \mathbb{F}_p}_{k+1 \text{ factors}}) \cong H_s(\underbrace{B\mathbb{Z}/p^\infty\mathbb{Z} \times \cdots \times B\mathbb{Z}/p^\infty\mathbb{Z}}_{k \text{ factors}}, \mathbb{F}_p)$$

and

$$\pi_s(T^k_B(B;C)) \cong H_s(\underbrace{BS^1 \times \cdots \times BS^1}_{k \text{ factors}}, \mathbb{F}_p).$$

It is easy to check that the induced homomorphism $\pi_s(T^k_A(A;C)) \to \pi_s(T^k_B(B;C))$ is (under the isomorphisms above) identified with $H_s(\underbrace{Bi \times \cdots \times Bi}_{k \text{ factors}}, \mathbb{F}_p)$, where $i:\mathbb{Z}/p^\infty\mathbb{Z} \to S^1$ is the inclusion. It is a well-known fact that this map is an isomorphism, which shows that the map in (8.4) above is an equivalence of spectra. In order to identify $\mathbb{H}(A)^\wedge_C$, then, it will suffice to identify $B^\wedge_C$. To do this, we consider the homomorphism of commutative S-algebras $g: S^0 \to \mathbb{H}(\mathbb{Z}) \to \mathbb{Z}[S^1]$. As before, let $\rho(\mathbb{Z}[S^1])$ denote $\mathbb{Z}[S^1]$ regarded as an S^0-

module via restriction of scalars along g. Then we obtain a natural map $\rho((\mathbb{Z}[S^1])^\wedge_{\mathbb{H}(\mathbb{F}_p)} \to \mathbb{Z}[S^1])^\wedge_{\mathbb{H}(\mathbb{F}_p)}$. Again, the left hand completion is over the commutative S-algebra S^0, and the right one is over the S-algebra $\mathbb{Z}$ $[S^1]$. There is a homomorphism of algebraic to geometric spectral sequences (7.1), which is an isomorphism since $\pi_0 S^0 \to \pi_0 \mathbb{Z}$ $[S^1]$. Consequently, the desired completion is the p-adic completion of the spectrum $\mathbb{Z}$ $[S^1]$. The resulting homotopy groups are the groups $H_i(S^1, Z_p)$, so we finally obtain the formula

$$l\pi_i(A^\wedge_{\mathbb{H}(\mathbb{F}_p)}) \cong \mathbb{Z}_p \quad \text{for } i = 0, 1$$

$$\pi_i(A^\wedge_{\mathbb{H}(\mathbb{F}_p)}) = 0 \quad \text{otherwise.}$$

This result has been extended to a statement about finitely generated nilpotent groups by T. Lawson in [13]. A construction referred to as deformation K-theory can be made which incorporates the topology on spaces of representations, and which applies to any discrete group. In the case of Z, it produces a spectrum with homotopy groups given by $\mathbb{Z}[S^1]) \otimes \pi_* ku$. What Lawson proves is that for a finitely generated nilpotent group Γ, the derived completion of the representation ring of the pro-p completion of Γ at the homomorphism given by mod p augmentation is equivalent to the p-adic completion of the deformation k-theory of Γ. This gives a link between this completion process on representation rings of pro-p quotients of Γ with the geometry of representation varieties of Γ.

ACKNOWLEDGMENTS

The research was supported in part by NSF DMS-0406992. The author would like to thank certain mathematicians with whom he had valuable conversations concerning this material. In particular, discussions with Bjørn Dundas, Bill Dwyer, John Greenlees, Mark Hovey, Rick Jardine, Mike Mandell, J. Peter May, Haynes Miller, and Brooke Shipley have been especially helpful.

REFERENCES

1. J.F. Adams, J.-P. Haeberly, S. Jackowski, J.P. May, A generalization of the Atiyah–Segal completion theorem, Topology 27 (1) (1988) 1–6.
2. M. Barr, J. Beck, Acyclic models and triples, in: 1966 Proc. Conf. Categorical Algebra (La Jolla, Calif., 1965), Springer, New York, pp. 336–343.
3. A.K. Bousfield, Homotopical localizations of spaces, Amer. J. Math. 119 (6) (1997) 1321–1354.
4. A.K. Bousfield, Cosimplicial Resolutions and Homotopy Spectral Sequences in model categories, in: Geometry and Topology, vol. 7, 2003, pp. 1001–1053.
5. A.K. Bousfield, D.M. Kan, Homotopy Limits, Completions and Localizations, in: Lecture Notes in Mathematics, vol. 304, Springer-Verlag, Berlin, New York, 1972, v+348 pp.
6. G. Carlsson, Recent Developments in Algebraic K-Theory, in: Current Developments in Mathematics, vol. 2001, International Press, 2002, pp. 1–40.
7. W.G. Dwyer, J. Spali´nski, Homotopy theories and model categories, in: Handbook of Algebraic Topology, North-Holland, Amsterdam, 1995, pp. 73–126.
8. W.G. Dwyer, J.P.C. Greenlees, S. Iyengar, Duality in algebra and topology, Adv. Math. 200 (2) (2006) 357–402.
9. D. Eisenbud, Commutative Algebra.With a View Toward Algebraic Geometry, in: Graduate Texts in Mathematics, vol. 150, Springer-Verlag, New York, 1995.
10. A.D. Elmendorf, I. Kriz, M.A. Mandell, J.P. May, Rings, Modules, and Algebras in Stable Homotopy Theory. With an Appendix by M. Cole, in: Mathematical Surveys and Monographs, vol. 47, American Mathematical Society, Providence, RI, 1997.
11. J.P.C. Greenlees, J.P. May, Completions in algebra and topology, in: Handbook of Algebraic Topology, North-Holland, Amsterdam, 1995, pp. 255–276.
12. M. Hovey, B. Shipley, J. Smith, Symmetric spectra, J. Amer. Math. Soc. 13 (1) (2000) 149–208.
13. T. Lawson, Completed representation ring spectra of nilpotent groups, Algebr. Geom. Topol. 6 (2006) 253–286.
14. M.A. Mandell, J.P. May, S. Schwede, B. Shipley, Model categories of diagram spectra, Proc. London Math. Soc. (3) 82 (2) (2001) 441–512.
15. D.G. Quillen, Finite generation of the groups Ki of rings of algebraic integers, in: Algebraic K-Theory, I: Higher K-Theories (Proc. Conf., Battelle Memorial Inst., Seattle, Wash., 1972), in: Lecture Notes in Math., vol. 341, Springer, Berlin, 1973, pp. 179–198.

16. A. Radulescu-Banu, Cofibrance and completion, Thesis, M.I.T. arXiv.org:math. AT/0612203, 1999.
17. S. Schwede, B.E. Shipley, Algebras and modules in monoidal model categories, Proc. London Math. Soc. (3) 80 (2000) 491–511.
18. B.E. Shipley, A convenient model category for commutative ring spectra, in: Homotopy Theory: Relations with Algebraic Geometry, Group Cohomology, and Algebraic K-Theory, in: Contemp. Math., vol. 346, Amer. Math. Soc., Providence, RI, 2004, pp. 473–483.
19. Weibel, A. Charles, An Introduction to Homological Algebra, in: Cambridge Studies in Advanced Mathematics, vol. 38, Cambridge University Press, Cambridge, ISBN: 0-521-43500-5, 1994, xiv+450 pp. ISBN: 0-521-55987-1.

CITATION

Gunnar Carlsson, Derived Completions in Stable Homotopy theory, Journal of Pure and Applied Algebra, Volume 212, Issue 3, March 2008, Pages 550-577, ISSN 0022-4049, http://dx.doi.org/10.1016/j.jpaa.2007.06.015.

Isomorphism Conjecture for Homotopy K-Theory and Groups Acting on Trees

Arthur Bartels and Wolfgang Lück

Fachbereich Mathematik, Universität Münster,
Einsteinstr. 62, 48149 Münster, Germany

ABSTRACT

We discuss an analogon to the Farrell–Jones Conjecture for homotopy algebraic K-theory. In particular, we prove that if a group G acts on a tree and all isotropy groups satisfy this conjecture, then G satisfies this conjecture. This result can be used to get rational injectivity results for the assembly map in the Farrell–Jones Conjecture in algebraic K-theory.

INTRODUCTION

The Farrell–Jones Conjecture [12] in algebraic K-theory is concerned with the K -theory $K_n(RG)$ of group rings RG for a group G and a ring R. The conjecture states that the assembly map

$$H_n^G(E_{\mathcal{VCYC}}(G); \mathbf{K}_R) \to K_n(RG) \tag{0.1}$$

is an isomorphism. (This map is constructed by applying a certain G -homology theory $H_n^G(-:,K_R)$ to the projection $E_{VCYC}(G) \to pt$, see Definition 1.1 and Remark 6.6.) There seem to occur two quite different phenomena in the algebraic K -theory of such group rings. Firstly, $K_n(RG)$ contains elements coming from the K-theory of RF for finite subgroups F of G . Secondly, it contains nilgroup information. This is already illuminated in the simple case G=Z, then $R[Z]=R[t,t^{-1}]$ and by the Bass–Heller–Swan formula [7] and [14]

$$Kn(R[Z]) \cong Kn(R) \oplus Kn\text{-}1(R) \oplus NKn(R) \oplus NKn(R) \tag{0.2}$$

Here, $NK_n(R)$ are the Nil-groups of R, which can be defined as the kernel of the projection $K_n(R[t]) \to K_n(R)$ induced from $t \mapsto 0$. In general, it is known [1] that the domain of the assembly map(0.1) splits as

$$H_n^G(E_{\mathscr{F}\mathscr{I}\mathscr{N}}(G); \mathbf{K}_R) \oplus H_n^G(E_{\mathscr{V}\mathscr{C}\mathscr{Y}\mathscr{C}}(G), E_{\mathscr{F}\mathscr{I}\mathscr{N}}(G); \mathbf{K}_R). \tag{0.3}$$

Thus, the Farrell–Jones Conjecture predicts a similar splitting for $K_n(RG)$.

In this paper, we will formulate a (Fibered) Isomorphism Conjecture for homotopy algebraic K-theory, seeConjecture 7.3. This variant of K-theory was defined by Weibel [31], building on the definition of Karoubi–Villamayor K-theory. The homotopy algebraic K-theory groups of a ring R are denoted by $KH_n(R)$. Their crucial property is homotopy invariance: $KH_n(R) \cong KH_n(R[t])$. In particular, homotopy algebraic K -theory does not contain Nil-groups. We think about this KH-Isomorphism Conjecture as an Isomorphism Conjecture for algebraic K-theory modulo Nil-groups. For a more precise formulation of the relation of the Farrell–Jones Conjecture in algebraic K -theory to the KH-Isomorphism Conjecture see Section 8.

Our main results concerning the KH-Isomorphism Conjecture are inheritance properties. A group G acts on a tree T, if T is a one-dimensional G-CW-complex which is contractible (after forgetting the group action).

Definition 0.4 the Class of Groups C_0.

We define the following properties a class C of groups may or may not have:

1. (FIN)All finite groups belong to C.
2. (TREE)Suppose that G acts on a tree T. Assume that for each $x \in T$ the isotropy group G_x belongs to C. Then G belongs to C.
3. (COL)Let G be a group with a directed system of subgroups $\{G_i | i \in I\}$, which is directed by inclusion and satisfies $\bigcup_{i \in I} G_i = G$. If each G_i belongs to C, then $G \in C$.
4. (SUB)If $G \in C$ and $H \subseteq G$ is a subgroup, then $H \in C$.
5. We define $_{C0}$ to be the smallest class of groups satisfying (FIN), (TREE) and (COL).

It is not hard to check that the class C_0 is closed under taking subgroups. For instance, if H is a subgroup of a group G acting on a tree, then H acts also on this tree and the isotropy groups satisfy $H_x \subseteq G_x$. By induction, we may assume that the G_x are closed under taking subgroups and therefore $H \in C_0$.

Theorem 0.5 Inheritance Properties of the KH-Isomorphism Conjecture
The class of groups satisfying the Fibered KH-Isomorphism Conjecture for a fixed coefficient ring R has the properties (FIN), (TREE), (COL) and (SUB). The class of groups satisfying the KH-Isomorphism Conjecture for a fixed coefficient ring R has the properties (FIN), (TREE) and (COL). In particular, all groups in C_0 satisfy the (Fibered) KH-Isomorphism Conjecture.

Remark 0.6
The class of groups satisfying the KH-Isomorphism Conjecture is strictly bigger than $_{C0}$ since it contains all fundamental groups of closed Riemannian manifolds with negative sectional curvature by [5] and Theorem 8.4(i).

This result has the following applications.

Theorem 0.7 Extensions of Groups and Actions on Trees

Let $1 \to K \to G \to Q \to 1$ be an extension of groups. Suppose that K acts on a tree with finite stabilizers and that Q satisfies the Fibered KH-Isomorphism Conjecture 7.3 for the ring R. Then G satisfies the Fibered KH-Isomorphism Conjecture 7.3 for the ring R.

A ring R is called regular if it is Noetherian and every finitely generated R-module possesses a finite-dimensional resolution by finitely generated projective modules.

Theorem 0.8 Conclusions for the K-Theoretic Farrell–Jones Conjecture for Groups in C

Let G be a group in the class C_0 defined above in (0.4). Then

i. Let R be a regular ring with $Q \subseteq R$. Then the assembly map

$$H_n^G(E_{\mathcal{F}\mathcal{I}\mathcal{N}}(G); \mathbf{K}_R) \to K_n(RG)$$

is injective, or, equivalently, the injectivity part of the Farrell–Jones Isomorphism Conjecture for algebraic K-theory is true for (G,R).

ii. Let R be the ring Z of integers. Then the assembly map

$$H_n^G(E_{\mathcal{F}\mathcal{I}\mathcal{N}}(G); \mathbf{K}_{\mathbb{Z}}) \to K_n(\mathbb{Z}G)$$

is rationally injective, or, equivalently, the rational injectivity part of the Farrell–Jones Isomorphism Conjecture for algebraic K-theory is true for (G,Z).

Proposition 0.9

The following classes of groups belong to C_0:

I. One relator groups.
II. G is poly-free, i.e. there is a filtration

$$\{1\} = G_0 \subseteq G_1 \subseteq G_2 \subseteq \cdots \subseteq G_n = G$$

such that G_i is normal in G_{i+1} with a free group as quotient G_{i+1}/G_i. The pure braid group is an example.

III. Let M be a compact orientable 3-manifold with prime decomposition $M = M_1 \# M_2 \# \cdots \# M_n$. Suppose that each M_i, which has infinite fundamental group and is aspherical, has a boundary or is a Haken manifold. Then $\pi_1(M) \in \mathcal{C}_{0}$.

IV. If M is a compact two-dimensional manifold, then $\pi_1(M) \in \mathcal{C}_{0}$.

V. If M is a submanifold of S^3, then $\pi_1(M) \in \mathcal{C}_{0}$.

Next, we discuss similar inheritance properties for the Farrell–Jones Conjecture in algebraic K-theory. A ring R is called regular coherent if every finitely presented R-module possesses a finite-dimensional resolution by finitely generated projective R-modules. A ring R is regular if and only if it is regular coherent and Noetherian. A group G is called regular or regular coherent, respectively, if for any regular ring R the group ring RG is regular, respectively regular coherent. For more information about these notions we refer to [29, Theorem 19.1].

Definition 0.10 the Classes of Groups CL and CL′

Consider the following further properties a class C of groups may have:

1. (TRI)The trivial group belongs to C.
2. (VCYC)All virtually cyclic groups belong to C.
3. (TREE$_R$)Suppose that G acts on a tree T. Assume that for each $x \in T$ the isotropy group G_x belongs to C. For each edge e of T, assume that the isotropy group G_e is regular coherent. Then G belongs to C.
4. The class CL is defined as the smallest class of groups satisfying (TRI), (TREE$_R$) and (COL). The classCL′ is defined as the smallest class of groups satisfying (VCYC), (TREE$_R$) and (COL).

All groups appearing in CL are torsionfree. Similar to the class C_0, the classes CL and CL′ are closed under taking subgroups [29, Proposition 19.3]. We conclude from Waldhausen [29, Theorem 17.5, p. 250]that CL contains a group G appearing in Proposition 0.9 under (i), (iii), (iv) and (v) provided that G is torsionfree. One of the main results in Waldhausen's article [29] is that for a regular ring R the K-theoretic assembly map

$$H_n(BG;K_R) \to K_n(RG)$$

is an isomorphism. Actually, Waldhausen states this only for $n \geqslant 0$, but the embedding of $_{Kn-1}(R)$ into $K_n(R[Z])$ allows the extension to all n, see for example Remark 10.3. Furthermore, Waldhausen considers HNN extensions and amalgamated products rather than action on trees, but this does not change the class CL, compare Remark 4.7 and Lemma 5.1.

Theorem 0.11

Let R be a regular ring. The class of groups satisfying the Farrell–Jones Conjecture in algebraic K-theory for the ring R has the properties (VCYC), (TREE$_R$)and (COL). In particular, all groups in CL satisfy the Farrell–Jones Conjecture in algebraic K-theory for the ring R.

Related results can be found in [15] and [24]. It is an interesting question, whether the class of groups satisfying the Farrell–Jones Conjecture in algebraic K-theory has the property (TREE).

Theorem 0.5 and Waldhausen's result imply

Corollary 0.12

Let R be a regular ring. Let G be a group in the class CL. Then the canonical map

$$K_n(RG) \to KH_n(RG)$$

is bijective for $n \in Z$.

Isomorphism Conjectures can be formulated in the quite general context of equivariant homology theories, see Definition 1.1. We show in Theorem 4.2 that the property (TREE) holds for the class of groups satisfying the Isomorphism Conjecture for such an equivariant homology theory whenever the equivariant homology theory satisfies the tree property, see Definition 4.1. The weaker property (TREE$_R$) is related to the regular tree property (see Definition 4.1), which is a weakening of the tree property. Theorem 0.7 has also an analogon in this setting, see Corollary 4.4. The tree property means essentially that there are Mayer–Vietoris sequences for amalgamated products and HNN extensions of groups in the equivariant homology theory (see Lemma 4.5). On spectrum level this means that there are certain homotopy cartesian diagrams of spectra (see Lemma 5.1 and Remark 5.6). In Section 7, we define the equivariant homology theory $H^?_*(-,KH_R)$ that is relevant for the KH-Isomorphism Conjecture. We prove inTheorem 11.1 that this theory satisfies the tree property. In the case of algebraic K-theory, amalgamated free products and HNN extensions have been analyzed by Waldhausen [29]. In both cases there are long exact sequences, but they involve as an additional term Waldhausen's Nil-groups. Their non-triviality obstructs the equivariant homology theory $H^?_*(-;K_R)$ relevant for the Farrell–Jones Conjecture in algebraic K-theory from having the tree property. Our proof of the tree property for $H^?_*(-,KH_R)$, consists essentially of showing that Waldhausen's Nil-groups are killed under the transition from K to KH. On the other hand, vanishing results for Waldhausen's Nil-groups can be used to show (see Theorem 11.4) that $H^?_*(-;K_R)$ has the regular tree property (see Definition 4.1). This implies then that for a regular ring R the class of groups satisfying the Farrell–Jones Conjecture has the property (TREE$_R$), seeTheorem 4.2(iii). It is an interesting question for which rings R the equivariant homology theory $H^?_*(-;K_R) \otimes \mathbb{Q}$ has the tree property. It is worthwhile to consider also $H^?_*(-;L_R^{-\infty})$, the equivariant homology theory relevant for the Farrell–Jones Conjecture in L-theory. In this case, amalgamated free products and HNN extensions have been analyzed by Cappell [9]. Again additional terms appear in the long exact sequences, the UNil-groups and non-triviality of those obstructs this theory from having the tree property. On the other hand, these UNil-groups are known to be

2-torsion [9], thus $H^?_n(-;L^{-\infty}_R) \otimes \mathbb{Z}[\frac{1}{2}]$ does have the tree property. Thus, we obtain the following result.

Theorem 0.13 Conclusions for the L -theoretic Farrell–Jones Conjecture for groups in C.

The class of groups for which the assembly map

$$H^G_n(E_{\mathcal{F}\mathcal{J}\mathcal{N}}(G); \mathbf{L}^{-\infty}_R) \to L^{-\infty}_n(RG)$$

becomes an isomorphism after tensoring with $\mathbb{Z}[\frac{1}{2}]$, has the properties (FIN), (TREE) and (COL). In particular, this class contains the class in C_0 from Definition 0.4.

In the context of topological K-theory, i.e. for the Baum–Connes Conjecture one can apply our results to the equivariant K -theory $H^G_n(-:,K^{top}) = K^G_n(-)$. Then one obtains the analogon of our Theorem 0.5. In this case, amalgamated products and HNN extensions have been analyzed Pimsner–Voiculescu [21] and Pimsner [20]. Here the situation is much better, since no Nil-groups appear. This analogon has already been proved by Oyono-Oyono [19] for the Baum–Connes Conjecture (with coefficients).

We are indebted to Holger Reich for pointing out Ref. [30] to us.

The paper is organized as follows:

1. Isomorphism Conjectures for equivariant homology theories
2. Homological aspects
3. Continuous equivariant homology theories
4. The tree property
5. Equivariant homology theories constructed from spectra
6. Isomorphism Conjectures for spectra
7. The KH-Isomorphism Conjecture
8. The Relation between the K - and the Kh-Isomorphism Conjecture
9. Non-connective Waldhausen Nil

ISOMORPHISM CONJECTURES FOR EQUIVARIANT HOMOLOGY THEORIES

We will use the notion of an equivariant homology theory $H_*^?$ with values in Λ-modules for a commutative associative ring Λ with unit from [16, Section 1]. This essentially means that we get for each group G a G-homology theory H_*^G which assigns to a (not necessarily proper or cocompact) pair of G-CW-complexes (X,A) a $\mathbb{Z}$-graded Λ-module $H_n^G(X,A)$, such that there exists natural long exact sequences of pairs and G-homotopy invariance, excision, and the disjoint union axiom are satisfied. Moreover, an induction structure is required which in particular implies for a subgroup $H\subseteq G$ and a H-CW-pair (X,A) that there is a natural isomorphism

$$\mathscr{H}_n^H(X, A) \xrightarrow{\cong} \mathscr{H}_n^G(G\times_H(X, A)).$$

We will later discuss examples, the most important ones will be given by those equivariant homology theories which appear in the Baum–Connes Conjecture and the Farrell–Jones Conjecture. These conjectures are special cases of the following more general formulation of a (Fibered) Isomorphism Conjecture (see Section 6).

A family F of subgroups of G is a set of subgroups which is closed under conjugation and taking subgroups. If C is a class of groups that is closed under taking subgroups and isomorphisms, then the collections of subgroups of G that are in C forms a family $C(G)$ of subgroups of G. Abusing notation, we will denote this family often by C. Examples are the families FIN of finite subgroups and VCYC of virtually cyclic subgroups. Given a group homomorphism $\varphi: K\to G$ and a family F of subgroups of G, define the family φ^*F of subgroups of

K by $\varphi^*F=\{H\subseteq K|\varphi(H)\in F\}$. If $i:H\to G$ is the inclusion of a subgroup, then we write often $F_{|H}$ for i^*F. Associated to such a family there is a G-CW -complex $E_F(G)$ (unique up to G -homotopy equivalence) with the property that the fix point sets $E_F(G)^H$ are contractible for $H\in F$ and empty for $H\notin F$. For F=ALL the family of all subgroups, we can take the one-point-space pt as a model for $E_{ALL}(G)$. For more information about the spaces we refer for instance to [17].

Definition 1.1 (Fibered) Isomorphism Conjecture for $H_*^?$

Let $H_*^?$ be an equivariant homology theory with values in Λ-modules. A group G together with a family of subgroups F satisfies the Isomorphism Conjecture (in the range $\leq N$) if the projection $pr:_{EF}(G)\to pt$ to the one-point-space pt induces an isomorphism

$$\mathscr{H}_n^G(pr):\mathscr{H}_n^G(E_{\mathscr{F}}(G))\stackrel{\cong}{\to}\mathscr{H}_n^G(pt)$$

for $n\in Z$ (with $n\leq N$).

The pair (G,F) satisfies the Fibered Isomorphism Conjecture (in the range $\leq N$) if for each group homomorphism $\varphi:K\to G$ the pair (K,φ^*F) satisfies the Isomorphism Conjecture (in the range $\leq N$).

Built in into the Fibered Isomorphism Conjecture is the following obvious inheritance property which is not known to be true in general in the non-fibered case.

Lemma 1.2

Let $H_*^?$ be an equivariant homology theory, let $\varphi:K\to G$ be a group homomorphism and let F be a family of subgroups. If (G,F) satisfies the Fibered Isomorphism Conjecture 1.1 (in the range $\leq N$), then (K,φ^*F) satisfies the Fibered Isomorphism Conjecture 1.1 (in the range $\leq N$).

Proof

If $\Psi: L \to K$ is a group homomorphism, then $\Psi^*(\varphi^*F)=(\varphi \circ \Psi)^*F$.

In particular, if for a given class of groups C, which is closed under isomorphism and taking subgroups, the Fibered Isomorphism Conjecture 1.1 is true for $(G,C(G))$ and if $H \subseteq G$ is a subgroup, then the Fibered Isomorphism Conjecture 1.1 is true for $(H,C(H))$.

HOMOLOGICAL ASPECTS

The disjoint union axiom ensures that a G-homology is compatible with directed colimits.

Lemma 2.1

Let H_*^G be a G-homology theory. Let X be a G-CW-complex and $\{X_i | i \in I\}$ be a directed system of G-CW-subcomplexes directed by inclusion such that $X = \bigcup_{i \in I} X_i$. Then for all $n \in Z$ the natural map

$$\operatorname*{colim}_{i \in I} \mathscr{H}_n^G(X_i) \overset{\cong}{\to} \mathscr{H}_n^G(X)$$

is bijective.

Proof
Compare for example with [27, Proposition 7.53, p. 121], where the non-equivariant case for I=N is treated. The main point is that the functor colimit over a directed system of R-modules is exact.

Lemma 2.2

Let $H_*^?$ be an equivariant homology theory with values in Λ-modules in the sense of [16, Section 1]. Let G be a group and let F be a family of subgroups of G. Let Z be a G-CW-complex. Consider $N \in Z \cup \{\infty\}$. Suppose for each $H \subseteq G$ which occurs as isotropy group in Z that the G-map induced by the projection $\mathrm{pr}: E_{F|H}(H) \to \mathrm{pt}$

$$\mathcal{H}_n^H(\mathrm{pr}): \mathcal{H}_n^H(E_{\mathcal{F}|H}(H)) \to \mathcal{H}_n^H(\mathrm{pt})$$

is bijective for all $n \in Z, n \leq N$.

Then the map induced by the projection $\mathrm{pr}_1: E_F(G) \times Z \to Z$

$$\mathcal{H}_n^G(\mathrm{pr}_1): \mathcal{H}_n^G(E_{\mathcal{F}}(G) \times Z) \to \mathcal{H}_n^G(Z)$$

is bijective for $n \in Z, n \leq N$.

Proof
We first prove the claim for finite-dimensional G-CW -complexes by induction over d=dim(Z). The induction beginning dim(Z)=-1, i.e. Z=Ø, is trivial. In the induction step from (d-1) to d we choose a G-pushout

$$
\begin{array}{ccc}
\coprod_{i \in I_d} G/H_i \times S^{d-1} & \longrightarrow & Z_{d-1} \\
\downarrow & & \downarrow \\
\coprod_{i \in I_d} G/H_i \times D^d & \longrightarrow & Z_d
\end{array}
$$

If we cross it with $E_F(G)$, we obtain another G-pushout of G-CW-complexes. The various projections induce a map from the Mayer–Vietoris sequence of the latter G-pushout to the Mayer–Vietoris sequence of the first G-pushout. By the Five-Lemma, it suffices to prove that the following maps

$$\mathcal{H}_n^G(\mathrm{pr}_2): \mathcal{H}_n^G\left(E_{\mathcal{F}}(G) \times \coprod_{i \in I_d} G/H_i \times S^{d-1}\right) \to \mathcal{H}_n^G\left(\coprod_{i \in I_d} G/H_i \times S^{d-1}\right),$$

$$\mathcal{H}_n^G(\mathrm{pr}_3): \mathcal{H}_n^G(E_{\mathcal{F}}(G) \times Z_{d-1}) \to \mathcal{H}_n^G(Z_{d-1}),$$

$$\mathcal{H}_n^G(\mathrm{pr}_4): \mathcal{H}_n^G\left(E_{\mathcal{F}}(G) \times \coprod_{i \in I_d} G/H_i \times D^n\right) \to \mathcal{H}_n^G\left(\coprod_{i \in I_d} G/H_i \times D^n\right).$$

are bijective for $n \in \mathbb{Z}$, $n \leq N$. This follows from the induction hypothesis for the first two maps. Because of the disjoint union axiom and G-homotopy invariance of $H_n^?$ the claim follows for the third map if we can show for any $H \subseteq G$ which occurs as isotropy group in Z that the map

$$\mathcal{H}_n^G(\mathrm{pr}_1): \mathcal{H}_n^G(E_{\mathcal{F}}(G) \times G/H) \to \mathcal{H}_n^G(G/H) \tag{2.3}$$

is bijective for $n \in \mathbb{Z}$, $n \leq N$. The G-map

$$G \times_H \mathrm{res}_G^H E_{\mathcal{F}}(G) \to G/H \times E_{\mathcal{F}}(G), \quad (g, x) \mapsto (gH, gx)$$

is a G-homeomorphism, where res_G^H denotes the restriction of the G-action to an H-action. Obviously, $\mathrm{res}_G^H E_{\mathcal{F}}(G)$ is a model for $E_{\mathcal{F}|H}(H)$. Since for any H-CW-complex Y there is a natural isomorphism $H_n^H(Y) \xrightarrow{\cong} H_n^G(G \times_H Y)$, the map (2.3) can be identified with the map

$$\mathcal{H}_n^G(\mathrm{pr}): \mathcal{H}_n^H(E_{\mathcal{F}|H}(H)) \to \mathcal{H}_n^H(\mathrm{pt})$$

which is bijective for all $n \in \mathbb{Z}$, $n \leq N$ by assumption. This finishes the proof in the case that Z is finite-dimensional.

Finally, we consider an arbitrary G-CW-complex Z. It can be written as the colimit $\mathrm{colim}_{d \to \infty} Z_d$. The natural maps

$$\mathrm{colim}_{d \to \infty} \mathcal{H}_n^G(E_{\mathcal{F}}(G) \times Z_d) \xrightarrow{\cong} \mathcal{H}_n^G(E_{\mathcal{F}}(G) \times Z),$$

$$\mathrm{colim}_{d \to \infty} \mathcal{H}_n^G(Z_d) \xrightarrow{\cong} \mathcal{H}_n^G(Z)$$

are bijective by Lemma 2.1. Since the colimit of isomorphisms is an isomorphism again, Lemma 2.2 follows.

Theorem 2.4 Reducing the Family

Let $H_*^?$ be an equivariant homology theory with values in Λ-modules. Let G be a group and let F$\subseteq$G be families of subgroups of G. Consider N$\in$Z$\cup\{\infty\}$. Suppose for each H$\in$G, or, more generally, suppose for each isotropy group appearing in a specific model for $E_G(G)$ that $(H, F_{|H})$ satisfies the (Fibered) Isomorphism Conjecture 1.1 (in the range $\leq$N).

Then (G, G) satisfies the (Fibered) Isomorphism Conjecture 1.1 (in the range $\leq$N) if and only if (G, F) satisfies the (Fibered) Isomorphism Conjecture 1.1 (in the range $\leq$N).

Proof
For the Isomorphism Conjecture this follows from Lemma 2.2 applied to the case $Z=E_G(G)$ and the fact that $E_F(G)\times E_G(G)$ is a model for $E_F(G)$. The case of the Fibered Isomorphism Conjecture is easily reduced to the former case.

Lemma 2.5

Let $H_*^?$ be an equivariant homology theory with values in Λ-modules. Let C be a class of groups that is closed under isomorphisms, subgroups and quotients. Let $1\to L \to G\overset{p}{\to} Q \to 1$ be an extension of groups. Suppose that $(Q;C(Q))$ satisfies the Fibered Isomorphism Conjecture 1.1 (in the range $\leq$N) and that for H$\in p^*C(Q)$ the pair $(H,C(H))$ satisfies the Fibered Isomorphism Conjecture 1.1 (in the range $\leq$N).

Then (G, C (G)) satisfies the Fibered Isomorphism Conjecture 1.1 (in the range $\leq$N).

Proof
By Lemma 1.2 the pair $(G,p^*C(Q))$ satisfies the Fibered Isomorphism Conjecture 1.1 (in the range $\leq$N). Since C is closed under quotients we have $C(G)\subseteq p^*C(Q)$. Now, the assumption on the subgroups H$\in p^*C(Q)$ and Theorem 2.4 imply the result.

CONTINUOUS EQUIVARIANT HOMOLOGY THEORIES

In this section, we explain a criterion for an equivariant homology theory ensuring that for a class C of groups closed under subgroups and isomorphisms the (Fibered) Isomorphism Conjecture 1.1 is true for $(G, C(G))$ provided that G is a directed union $G = \bigcup_{i \in I} G_i$ of groups G_i and the (Fibered) C-Isomorphism Conjecture 1.1 is true for $(G_i, C(G_i))$ for all $i \in I$.

Definition 3.1 Continuous Equivariant Homology Theory

An equivariant homology theory $H_*^?$ is called continuous if for each group G and directed system of subgroups $\{G_i | i \in I\}$, which is directed by inclusion and satisfies $\bigcup_{i \in I} G_i = G$, and each $n \in Z$ the map

$$\operatorname*{colim}_{i \in I} j_i : \operatorname*{colim}_{i \in I} \mathscr{H}_n^{G_i}(\mathrm{pt}) \to \mathscr{H}_n^G(\mathrm{pt})$$

is an isomorphism, where $j_i : H_n^G(G/G_i)$ is the composition of the induction isomorphism $H_n^{G_i}(\mathrm{pt}) \xrightarrow{\cong} H_n^G(G/G_i)$ with the map induced by the projection $G/G_i \to \mathrm{pt}$.

Lemma 3.2

Let $H_*^?$ be a continuous equivariant homology theory. Let G be a group with a directed system of subgroups $\{G_i | i \in I\}$, which is directed by inclusion and satisfies $\bigcup_{i \in I} G_i = G$.

Then for each G-CW-complex X and each $n \in Z$ the map

$$\operatorname*{colim}_{i \in I} j_i : \operatorname*{colim}_{i \in I} \mathscr{H}_n^{G_i}(\mathrm{res}_G^{G_i} X) \to \mathscr{H}_n^G(X)$$

is an isomorphism, where $j_i : H_n^{G_i}(\mathrm{res}_G^{G_i} X) \to H_n^G(X)$ is the composition of the induction isomorphism $H_n^{G_i}(\mathrm{res}_G^{G_i} X) \xrightarrow{\cong} H_n^G(G \times_{G_i} \mathrm{res}_G^{G_i} X)$ with the homomorphism induced by the G-map $G \times_{G_i} \mathrm{res}_G^{G_i} X \to X$ that sends (g,x) to gx.

Proof

Since $\mathrm{colim}_{i\in I}$ is an exact functor, $\mathrm{colim}_{i\in I} H_n^{G_i}(\mathrm{res}_G^{G_i} X)$ is a G-homology theory in X. The map $\mathrm{colim}_{i\in I} ji$ is a transformation of G-homology theories. Therefore, it suffices to prove that

$$\mathrm{colim}_{i\in I} \; j_i : \mathrm{colim}_{i\in I} \; \mathscr{H}_n^{G_i}(\mathrm{res}_G^{G_i} G/H) \to \mathscr{H}_n^G(G/H)$$

is an isomorphism for every subgroup $H \subseteq G$ and $n \in Z$.

For $i \in I$ let $k_i : G_i G_i \cap H \to \mathrm{res}_G^{G_i} G/G_i$ be the obvious injective G_i map. Then the following diagram commutes

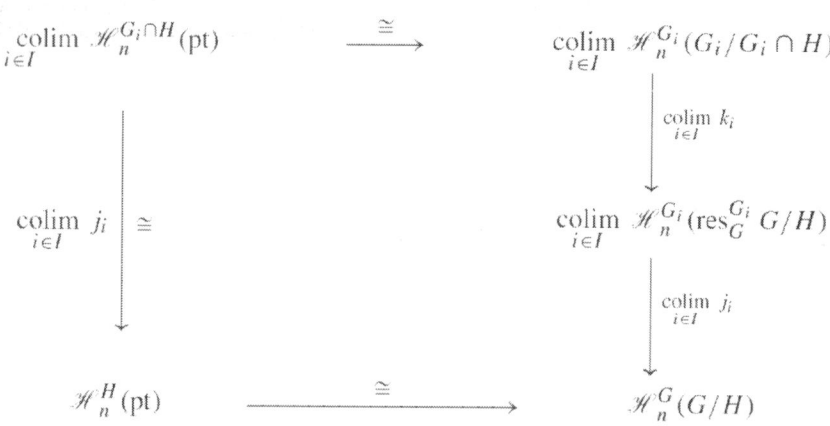

where the horizontal maps are the isomorphism given by induction. The left vertical arrow is bijective since $H_*^?$ is continuous by assumption. Hence, it remains to show that the map

$$\operatorname*{colim}_{i \in I} k_i \colon \operatorname*{colim}_{i \in I} \mathscr{H}_n^{G_i}(G_i/G_i \cap H) \to \operatorname*{colim}_{i \in I} \mathscr{H}_n^{G_i}(\operatorname{res}_G^{G_i} G/H) \qquad (3.3)$$

is surjective.

We get an obvious decomposition of G_i-sets

$$\operatorname{res}_G^{G_i} G/H = \coprod_{G_i g H \in G_i \backslash G/H} G_i/G_i \cap g H g^{-1}$$

It induces an identification

$$\mathscr{H}_n^{G_i}(\operatorname{res}_G^{G_i} G/H) = \bigoplus_{G_i g H \in G_i \backslash G/H} \mathscr{H}_n^{G_i}(G_i/G_i \cap g H g^{-1}).$$

The summand corresponding to $_{G_i}1H$ is precisely the image of

$$\mathscr{H}_n^{G_i}(k_i) \colon \mathscr{H}_n^{G_i}(G_i/G_i \cap H) \to \mathscr{H}_n^{G_i}(\operatorname{res}_G^{G_i} G/H).$$

Consider an element $G_i g H \in G_i \backslash G/H$. Choose an index j with $j \geq i$ and $g \in G_j$. Then the structure map for $i \leq j$ is a map $H_n^{G_i}(\operatorname{res}_G^{G_i} G/H) \to H_n^{G_j}(\operatorname{res}_G^{G_j} G/H)$ which sends the summand corresponding to $G_i g H \in G_i \backslash G/H$ to the image of

$$\mathscr{H}_n^{G_j}(k_j) \colon \mathscr{H}_n^{G_j}(G_j/G_j \cap H) \to \mathscr{H}_n^{G_j}(\operatorname{res}_G^{G_j} G/H).$$

This implies that the map (3.3) is surjective. This finishes the proof of Lemma 3.2.

Proposition 3.4

Let $H_*^?$ be an equivariant homology theory which is continuous. Let C be a class of groups that is closed under isomorphism and taking subgroups. Let G be the directed union $G = \bigcup_{i \in I} G_i$ of subgroups G_i such that the (Fibered) Isomorphism Conjecture 1.1 (in the range $\leq N$) is true for $(G_i, C(G_i))$ for all $i \in I$.

Then the (Fibered) Isomorphism Conjecture 1.1 (in the range $\leq N$) is true for $(G, C(G))$.

Proof

Since $H_*^?$ is continuous by assumption, we get the isomorphism

$$\operatorname{colim}_{i \in I} \mathcal{H}_*^{G_i}(\mathrm{pt}) = \mathcal{H}_*^G(\mathrm{pt})$$

and from Lemma 3.2 the isomorphism

$$\mathcal{H}_*^G(E_{\mathcal{C}(G)}(G)) = \operatorname*{colim}_{i \in I} \mathcal{H}_*^{G_i}(\operatorname{res}_G^{G_i} E_{\mathcal{C}(G)}(G)).$$

The result follows for the Isomorphism Conjecture since the colimit of an isomorphism is an isomorphism and since $\operatorname{res}_G^{G_i} E_{C(G)}(G)$ is a model for $E_{C(G_i)}(G_i)$. If $\varphi: K \to G$ is a group homomorphism then the same argument can be applied to the triple $(K, \varphi^*C(G), \{\varphi^{-1}(G_i) | i \in I\})$ in place of $(G, C(G), \{G_i | i \in I\})$ and this implies the statement for the Fibered Isomorphism Conjecture.

THE TREE PROPERTY

In this section, we study criteria for an equivariant homology theory that ensure that the class of groups G for which (G,FIN) satisfies the (Fibered) Isomorphism Conjecture 1.1 has property (TREE) fromDefinition 0.4 or that the class of groups G for which (G,VCYC) satisfies the Isomorphism Conjecture 1.1 has property (TREE$_R$) from Definition 0.10.

Definition 4.1 Tree Property

An equivariant homology theory $H_*^?$ has the tree property if for any group G that acts on a tree T, the projection $\mathrm{pr}: T \to \mathrm{pt}$ induces for all $n \in \mathbb{Z}$ (with $n \leq N$) isomorphisms

$$\mathcal{H}_n^G(\mathrm{pr}): \mathcal{H}_n^G(T) \to \mathcal{H}_n^G(\mathrm{pt}).$$

It has the regular tree property if for any group G that acts on a tree T, such that for each edge e of T the isotropy group G_e is regular coherent, the projection pr:T→pt induces for all n∈Z (with n≤N) isomorphisms

$$\mathcal{H}_n^G(\mathrm{pr}): \mathcal{H}_n^G(T) \to \mathcal{H}_n^G(\mathrm{pt}).$$

Theorem 4.2 the Tree Property and Inheritance Properties of Isomorphism Conjectures

Let $H_*^?$ be an equivariant homology theory. Let Cbe a class of groups closed under subgroups and isomorphisms. Let $D_{fib}(C)$be the class of groups G for which Fibered Isomorphism Conjecture 1.1 (in the range ≤N) is true for (G,C(G))and let D(C)be the class of groups G for which Isomorphism Conjecture1.1 (in the range ≤N) is true for (G,C(G)).

I. Suppose that $H_*^?$ has the tree property (4.1). Then the class $D_{fib}(C)$ has the property (TREE) from Definition 0.4.

II. Suppose $H_*^?$ has the tree property (4.1) and C⊆FIN. Then the class D(C) has the property (TREE) from Definition 0.4.

III. Suppose that $H_*^?$ has the regular tree property (4.1) and C⊆VCYC. Then the class D(C) has the property (TREE$_R$) from Definition 0.10.

Proof

Let G act on a tree T. Denote by V the set of vertices of T and by E the set of edges. For x∈V∪E denoted by G_x the isotropy group of x and by $\varphi_x:G_x \to G$ the inclusion. Let $I_T=\{H \leqslant G | T^H \neq \emptyset\}$. Since in a tree there is a unique geodesic between any two points, the fixed set T^H is contractible for H∈I_T. Thus, T is a model for $E_{I_T}(G)$.

Next, we prove (i). In this case, we assume that for each x∈V∪E the pair $(G_x, C(G_x))$ satisfies the Fibered Isomorphism Conjecture 1.1. Let $\varphi:K \to G$ be a group homomorphism. Then K acts via φ on T. Equipped with this action T is also a model for $E_{\varphi^*I_T}(K)$. The tree property implies that (K,φ^*I_T) satisfies the Isomorphism Conjecture 1.1. Thus, (G,I_T) satisfies the Fibered Isomorphism Conjecture 1.1. Since the isotropy groups of T satisfy the Fibered Isomorphism Conjecture 1.1 with respect to C, we can deduce from Theorem 2.4 that $(G, C(G) \cap I_T)$ satisfies the Fibered Isomorphism Conjecture 1.1. Finally, we use the fact that for the Fibered Isomorphism Conjecture 1.1 we can always enlarge the family (see [4, Lemma 1.6]) to conclude that the pair $(G, C(G))$ satisfies the Fibered Isomorphism Conjecture 1.1.

Next, we prove (ii). In this case, we assume that for each $x \in V \cup E$ the pair $(G_x, C(G_x))$ satisfies the Isomorphism Conjecture 1.1. Arguing as above, we conclude that $(G, I_T \cap C(G))$ satisfies the Isomorphism Conjecture 1.1. Finite groups cannot act without fixed points on trees [26, Theorem 15, 6.1, p. 58 and 6.3.1, p. 60]. Therefore, $I_T \cap C(G) = C(G)$.

Finally, we prove (iii). In this case, we assume that for each $x \in V \cup E$ the pair $(G_x, C(G_x))$ satisfies the Isomorphism Conjecture 1.1 and that G_e is regular coherent for each $e \in E$. Arguing as before, we conclude that $(G, I_T \cap C(G))$ satisfies the Isomorphism Conjecture 1.1. We have to show that $(G, C(G))$ satisfies the Isomorphism Conjecture 1.1. Because of Theorem 2.4 it suffices to show for any virtually cyclic group $V \in C(G)$ that the Isomorphism Conjecture 1.1 holds for $(V, I_T \cap C(G)_{|V}) = (V, I_T)$.

We first consider the case, where V contains a non-trivial normal finite subgroup F. We saw above that T^F is not empty and contractible. By Lemma 4.3, regular coherent groups are torsion free. Thus, isotropy groups of edges are torsion free, therefore T^F is just a single vertex of T. Since F is normal in V, the action of V leaves the fixed points of F invariant. Therefore, the vertex T^F is a fixed point for V. Hence, we have $V \in I_T$ so that I_T consists of all subgroups of V.

If V does not contain a non-trivial normal finite subgroup F, it is either Z or the infinite dihedral group. In both cases, V acts on the tree R with finite stabilizers such that the stabilizers of the edges are trivial and every finite subgroup of V occurs as stabilizer. The tree R is a model for $E_{FIN}(V)$. Since $H^?_*$ has the regular tree property (4.1) the map

$H_n^V(\mathbb{R}) \to H_n^V(\text{pt})$ is bijective for all $n \in Z$. This shows that V satisfies the Isomorphism Conjecture 1.1 for (V, FIN(V)). If V=Z, then every subgroup $H \subseteq V$ is trivial or isomorphic to Z. If V is the infinite dihedral group, then any subgroup H of V is finite, infinite cyclic or infinite dihedral. We conclude from Theorem 2.4 that V satisfies the Isomorphism Conjecture 1.1 for every family which contains FIN, in particular for I_T.

Lemma 4.3
Regular coherent groups are torsion free.

Proof
Assume that F is a finite subgroup of a regular coherent group G. Then the ZG-module Z [G/F] is finitely presented and has a finite-dimensional resolution by finitely generated projective ZG-modules since G is regular coherent and the ring Z is regular. Thus, the restriction of Z [G/F] to a ZF-module has a finite-dimensional resolution by projective (but no longer finite generated) ZF-modules. As an ZF-module Z [G/F] contains Z (with the constant F -action) as a direct summand. Therefore, Z has a finite-dimensional resolution by projective ZF-modules. This is only possible if F is the trivial group.

Corollary 4.4

Let $H_*^?$ be an equivariant homology theory which has the tree property (see Definition 4.1). Let $1 \to K \to G \to Q \to 1$ be an extension of groups. Suppose that K acts on a tree with finite stabilizers and that (Q, FIN) satisfies the Fibered Isomorphism Conjecture 1.1 (in the range $\leq N$).

Then (G, FIN) satisfies the Fibered Isomorphism Conjecture 1.1 (in the range $\leq N$).

Proof
We first treat the case Q= {1}. Then the claim follows from Theorem 4.2(i) because for a finite group F the pair (F, FIN) obviously satisfies the Fibered Isomorphism Conjecture 1.1.

Next, we treat the case, where Q is finite. By a result of Dunwoody [11, Theorem 1.1] a group K acts on a tree with finite stabilizers if and only if $H^p(K;Q)=0$ for each $p \geq 2$. Since K acts on a tree with finite stabilizers, the trivial QK-module Q has a one-dimensional projective resolution. Hence, the trivial QG-module Q has a one-dimensional projective resolution since [G:K] is finite and invertible in Q. This implies $H^p(G;Q)=0$ for each $p \geqslant 2$. Hence, also G acts on a tree with finite stabilizers if Q is finite. This proves the claim for finite Q.

Now, the general case follows from Lemma 2.5.

Lemma 4.5

Let $H^?_*$ be an equivariant homology theory which is continuous. Then the following assertions are equivalent.

I. For each one-dimensional G-CW-complex T for which each component is contractible (after forgetting the group action), the projection $p_{rT}:T \to \pi_0(T)$ induces isomorphisms \

$$\mathcal{H}^G_n(\mathrm{pr}_T): \mathcal{H}^G_n(T) \xrightarrow{\cong} \mathcal{H}^G_n(\pi_0(T)),$$

for each $n \in Z$, where we consider $\pi_0(T)$ as a G-space using the discrete topology.

II. $H^?_*$ has the tree property, i.e. for each one-dimensional G-CW-complex T, which is contractible (after forgetting the group action), and each $n \in Z$ we obtain isomorphisms

$$\mathcal{H}^G_n(\mathrm{pr}_X): \mathcal{H}^G_n(X) \xrightarrow{\cong} \mathcal{H}^G_n(\mathrm{pt}).$$

III. For each one-dimensional G-CW-complex X, which is contractible (after forgetting the group action) and has only one equivariant 1-cell, and each $n \in Z$ we obtain isomorphisms

$$\mathcal{H}^G_n(\mathrm{pr}_X): \mathcal{H}^G_n(X) \xrightarrow{\cong} \mathcal{H}^G_n(\mathrm{pt}).$$

These three assertions remain equivalent if we add the requirement that the isotropy groups of edges are regular coherent to each assertion. (Thus, (ii) becomes the assertion that $H_*^?$ has the regular tree property.)

Proof

(i) $\Rightarrow$ (ii) $\Rightarrow$ (iii) is obvious.

(iii) $\Rightarrow$ (i): We prove the claim first under the assumption that $G\backslash T$ has finitely many 1-cells.

We use induction over the number of 1-cells in $G\backslash T$. In the induction beginning, where $G\backslash T$ has no 1-cell, T is the disjoint union of homogeneous spaces and the claim follows from the fact that $H_*^?$ satisfies the disjoint union axiom.

In the induction step, we can write T as a G-pushout

$$
\begin{array}{ccc}
G/H \times S^0 & \xrightarrow{\;q\;} & T_0 \\
\downarrow & & \downarrow \\
G/H \times D^1 & \longrightarrow & T
\end{array}
$$

for a G-CW -subcomplex $T_0 \subseteq T$ such that $G\backslash T_0$ has one 1-cell less than $G\backslash T$. Here, H is the isotropy group of the 1-cell of T that is not contained in T_0. Since a connected subgraph of a tree is again a tree, each component of T_0 is contractible. The induction hypothesis applies to T_0, $G/H \times S^0$ and $G/H \times D^1$. Define X to be the G-pushout

$$
\begin{array}{ccc}
G/H \times S^0 & \xrightarrow{\mathrm{pr}_{T_0} \circ q} & \pi_0(T_0) \\
\downarrow & & \downarrow \\
G/H \times D^1 & \longrightarrow & X
\end{array}
$$

The G-maps $p_{rT0}:T_0 \to \pi_0(T_0)$, $\mathrm{id}_{G/H \times S^0}$ and $\mathrm{id}_{G/H \times}D_1$ are non-equivariant homotopy equivalences and induce a G-map $f:T \to X$ which is a non-equivariant homotopy equivalence since $G/H \times S^0 \to G/H \times D^1$ is a cofibration. In particular, X is a one-dimensional G-CW-complex whose components are contractible. By a Mayer–Vietoris argument and the Five-Lemma the map

$$\mathcal{H}_n^G(f): \mathcal{H}_n^G(T) \xrightarrow{\cong} \mathcal{H}_n^G(X)$$

is bijective for all $n \in \mathbb{Z}$. The following diagram commutes:

$$
\begin{array}{ccc}
T & \xrightarrow{\mathrm{pr}_T} & \pi_0(T) \\
\downarrow{f} & & \downarrow{\pi_0(f)} \\
X & \xrightarrow[\mathrm{pr}_X]{} & \pi_0(X)
\end{array}
$$

Since the map $\pi_0(f)$ is bijective and hence a G-homeomorphism, $H_n^G(\pi 0(f))$ is bijective for all $n \in \mathbb{Z}$. Recall that we have to show that $H_n^G(\mathrm{pr}_T)$ is bijective for all $n \in \mathbb{Z}$. Hence, it suffices to show that $H_n^G(\mathrm{pr}_X)$ is bijective for all $n \in \mathbb{Z}$. This follows from the fact that we can write X as a disjoint union of a G-CW-complex Y, for which the assumption (iii) applies, and a zero-dimensional G-CW-complex Z, for which the induction beginning applies, and that $H_*^?$ satisfies the disjoint union axiom.

Next, we treat the general case. Because $H_*^?$ satisfies the disjoint union axiom, we can assume without loss of generality that $G\backslash T$ is connected. Since we can write $T = G_{\times H}T'$ for a path component T' and we have natural isomorphisms $H^H(T') \xrightarrow{\cong} H_n^G(T)$ and $H^H(\mathrm{pt}) \xrightarrow{\cong} H_n^G(G/H)$, we can assume without loss of generality that T is contractible.

Fix a 0-cell $e \in G\backslash T$. Let I be the set of finite connected CW-subcomplexes $Z \subseteq G\backslash T$ with $e \in Z$. It can be directed by inclusion and satisfies

$G/T = \bigcup_{Z \in I} Z$. Let $p: T \to G \backslash T$ be the projection. Then T is the directed union of the G-CW -subcomplexes $p^{-1}(Z)$. Because of Lemma 2.1 the canonical map

$$\operatorname*{colim}_{Z \in I} \mathcal{H}_n^G(p^{-1}(Z)) \xrightarrow{\cong} \mathcal{H}_n^G(T)$$

is bijective. Since each G-CW -complex $p^{-1}(Z)$ has only finitely many equivariant 1-cells and hence satisfies the claim, and a colimit of a system of isomorphisms is again an isomorphism, it suffices to show that

$$\operatorname*{colim}_{Z \in I} \mathcal{H}_n^G(\pi_0(p^{-1}(Z))) \to \mathcal{H}_n^G(\mathrm{pt}) \tag{4.6}$$

is bijective. Fix $(\tilde{e}) \in T$ with $p(\tilde{e}) = e$. Let G_Z be the isotropy group of the path component of $p^{-1}(Z)$ containing $\tilde{e}$ in the G -set $\pi_0(p^{-1}(Z))$. Since each Z is connected, $\pi_0(p^{-1}Z)$ is G/G_Z. We have for any inclusion $Z_1 \subseteq Z_2$ for elements $Z_1, Z_2 \in I$, that G_{Z_1} is a subgroup of G_{Z_2}. We have $G = \bigcup_{Z \in I} G_Z$. Since $H_*^?$ is continuous, we get an isomorphism

$$\operatorname*{colim}_{Z \in I} \mathcal{H}_n^G(G/G_Z) \xrightarrow{\cong} \mathcal{H}_n^G(\mathrm{pt}).$$

But this isomorphism can easily be identified with the map (4.6). This finishes the proof of Lemma 4.5.

Remark 4.7

Let G act on a tree T, such that $G \backslash T$ has only finitely many 1-cells. The proof of Lemma 4.5 shows that then G acts on tree X with the following properties: the quotient $G \backslash X$ has only one 1-cell. For each edge eof X the isotropy group G_e is also the isotropy group of an edge e' of T. For each vertex v of X there is a subtree T_v of T that is invariant under the isotropy group G_v and for which $G_v \backslash T_v$ has one less 1-cell than $G \backslash T$. In combination with the colimit argument from the proof of Lemma 4.5

this means that a class of groups C that has property (COL) from Definition 0.4 has property (TREE) from Definition 0.4 if and only if it has the following property:

(TREE′)Suppose that G acts on a tree T, where T has only one equivariant 1-cell. Assume that for each x∈Tthe isotropy group G_x belongs to C. Then G belongs to C;

and has property (TREE$_R$) if and only if it has the property:

((TREE$_R'$)) Suppose that G acts on a tree T. Assume that for each x∈T the isotropy group G_x belongs to C. For each edge e of T, assume that the isotropy group G_e is regular coherent. Then G belongs to C.

Note on the other hand, that the statement that the Fibered Isomorphism Conjecture 1.1 has property (TREE′) is really a statement about arbitrary actions on trees: if G acts on a tree T, where T has only one equivariant 1-cell and φ: K→G is a group homomorphism, then the induced action of K on T may have more equivariant 1-cells and may even be no longer cocompact. Therefore, we have to consider general trees in the formulation of the tree property in Definition 4.1.

EQUIVARIANT HOMOLOGY THEORIES CONSTRUCTED FROM SPECTRA

In this section, we want to give a criterion when an equivariant homology theory has the tree property provided that it arises from a covariant functor E:GROUPOIDS→SPECTRA which sends equivalences of groupoids to weak equivalences of spectra. This will be the main example for us.

Fix a group G. The transport groupoid $G^G(S)$ of a G-set S has S as set of objects and the set of morphism from s_1 to s_2 consists of those element g∈G with $g_{s1}=s_2$. Composition of morphisms comes from the group structure on G. The orbit category Or (G) has as objects homogeneous spaces G/H and as morphisms G -maps. We obtain a covariant functor

G^G: Or $(G) \rightarrow$ GROUPOIDS, $G/H \mapsto G^G(G/H)$. Define the covariant functor E^G: Or $(G) \rightarrow$ SPECTRA by $E \circ G^G$. Let $H_*^G(-;E)$ be the G-homology theory associated to E^G in [10, Sections 4, 7]. It is not hard to construct the relevant induction structure to get an equivariant homology theory $H_*^?(-;E)$. It has the property that for each group G with subgroup $H \subseteq G$ and each $n \in Z$ we have canonical isomorphisms

$$H_n^G(G/H; \mathbf{E}) \cong H_n^H(\text{pt}; \mathbf{E}) \cong \pi_n(\mathbf{E}(H)).$$

In the expression $E(H)$, we think of the group H as a groupoid with one object. More details of the construction of $H_*^?(-;E)$ can be found in [18] and [25].

Lemma 5.1

The equivariant homology theory $H_*^?(-,E)$ is continuous and has the tree property if and only if the following conditions are satisfied:

i. For each group G and directed system of subgroups $\{G_i | i \in I\}$, which is directed by inclusion and satisfies $\bigcup_{i \in I} G_i = G$, and each $n \in Z$ the map

$$\operatorname*{colim}_{i \in I} j_i : \operatorname*{colim}_{i \in I} \pi_n(\mathbf{E}(G_i)) \rightarrow \pi_n(\mathbf{E}(G))$$

is an isomorphism, where j_i is the homomorphism induced by the inclusion $G_i \rightarrow G$.

ii. Consider a pushout of groups

$$
\begin{array}{ccc}
H_0 & \xrightarrow{\;i_1\;} & H_1 \\
\downarrow{\scriptstyle i_2} & & \downarrow{\scriptstyle j_1} \\
H_2 & \xrightarrow[\;j_2\;]{} & G
\end{array}
$$

(5.2)

such that i_1 and i_2 are injective. In other words, G is the amalgamated product of H_1 and H_2 over H_0 with respect to the injections i_1 and i_2. Then for each such pushout (5.2) the following square of spectra is homotopy cocartesian:

$$
\begin{array}{ccc}
\mathbf{E}^G(G/H_0) \vee \mathbf{E}^G(G/H_0) & \xrightarrow{\mathbf{E}^G(\mathrm{pr}_1) \vee \mathbf{E}^G(\mathrm{pr}_2)} & \mathbf{E}^G(G/H_1) \vee \mathbf{E}^G(G/H_2) \\
{\scriptstyle \mathrm{id} \vee \mathrm{id}} \downarrow & & \downarrow {\scriptstyle \mathbf{E}^G(\mathrm{pr}_3) \vee \mathbf{E}^G(\mathrm{pr}_4)} \\
\mathbf{E}^G(G/H_0) & \xrightarrow[\mathbf{E}^G(\mathrm{pr}_5)]{} & \mathbf{E}^G(G/G)
\end{array}
$$

(5.3)

where the maps labeled p_{ri} denote canonical projections.

iii. Let $i_0, i_1 : H \to K$ be injective group homomorphisms. Let G be the HNN extension associated to i_0 and i_1. The HNN extension comes with an inclusion $j : K \to G$ and $t \in G$ such that $j \circ_{i0} = {}_{ct} \circ j \circ_{i1}$, where c_t is conjugation by t. (This is the defining property of the HNN extension.) We will use i_0 to consider H as a subgroup of G. Then the following square of spectra is homotopy cocartesian:

$$
\begin{array}{ccc}
\mathbf{E}^G(G/H) \vee \mathbf{E}^G(G/H) & \xrightarrow{\mathbf{E}^G(\mathrm{pr}_0) \vee \mathbf{E}^G(\beta)} & \mathbf{E}^G(G/K) \\
{\scriptstyle \mathrm{id} \vee \mathrm{id}} \downarrow & & \downarrow {\scriptstyle \mathbf{E}^G(\mathrm{pr}_1)} \\
\mathbf{E}^G(G/H) & \xrightarrow[\mathbf{E}^G(\mathrm{pr}_2)]{} & \mathbf{E}^G(G/G)
\end{array}
$$

(5.4)

where the maps labeled p_{ri} are canonical projections while β is defined by $\beta(gH) = gtK$.

For conditions (ii) and (iii) one can also consider the regular versions where (5.3) is only required to be homotopy cartesian if in addition H_0 is regular coherent and torsionfree and (5.4) is only required to be homotopy cartesian if in addition H is regular coherent and torsion-free. Then the equivariant homology theory $H_*^?(-, E)$ is continuous and has the regular tree property if and only the condition (i) and the regular versions of conditions (ii) and (iii) hold.

Proof

Obviously, condition (i) is equivalent to the condition that $H_*^?$ is continuous. From now on we assume that $H_*^?$ is continuous.

Suppose that the two conditions (ii) and (iii) are satisfied. Because of Lemma 4.5 it suffices to prove the tree property only for one-dimensional contractible G-CW-complexes T such that there is precisely one equivariant 1-cell. Such a G-CW-complex will have precisely one or precisely two equivariant 0-cells. We only treat the case, where there are two equivariant 0-cells, the proof of the other case is analogous using condition (iii) instead of condition (ii).

We can write T as a G-pushout

$$
\begin{array}{ccc}
G/H_0 \times S^0 & \xrightarrow{\;\mathrm{pr}_1 \,\amalg\, \mathrm{pr}_2\;} & G/H_1 \amalg G/H_2 \\
\downarrow & & \downarrow \\
G/H_0 \times D^1 & \longrightarrow & T
\end{array}
$$

where H_0 is a subgroup of both H_1 and H_2 and p_{r1} and p_{r2} are the canonical projections. Recall that a G-space Z defines a contravariant functor $\mathrm{Or}(G) \to \mathrm{SPACES}, G/H \mapsto \mathrm{map}_G(G/H, Z)$ and that we get a spectrum $\mathrm{map}_G(G/?, Z) \wedge_{\mathrm{Or}(G)} \mathbf{E}^G$ by the tensor product over the orbit category (see [10, Section 1]). If we apply $\mathrm{map}_G(G/?, -) \wedge_{\mathrm{Or}(G)} \mathbf{E}^G$ to the G-pushout ab006Fve, we obtain a homotopy cocartesian diagram of spectra

$$
\begin{array}{ccc}
\mathbf{E}^G(G/H_0) \vee \mathbf{E}^G(G/H_0) & \xrightarrow{\;\mathbf{E}^G(\mathrm{pr}_1) \vee \mathbf{E}(\mathrm{pr}_2)\;} & \mathbf{E}^G(G/H_1) \vee \mathbf{E}^G(G/H_2) \\
\mathrm{id}\vee\mathrm{id}\downarrow & & \downarrow \\
\mathbf{E}^G(G/H_0) & \longrightarrow & \mathrm{map}_G(G/?, T) \wedge_{\mathrm{Or}(G)} \mathbf{E}^G
\end{array}
\tag{5.5}
$$

The following diagram is a pushout of groups

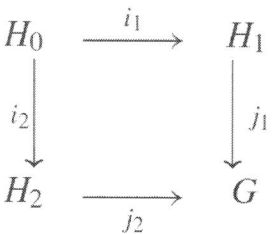

where $i_k : H_0 \to H_k$, $j_k : H_k \to G$ are inclusion (see [26, Example 1, p. 43]). Hence by condition (ii), we have the homotopy cocartesian square (5.3). The projection $\mathrm{pr} : T \to G/G$ induces a map from the right lower corner of the diagram (5.5) to the right lower corner of the diagram (5.3), if we identify

$G/G \wedge_{\mathrm{Or}(G)} E^G = E^G(G/G)$. If we take the identity on the other three corners, we get a map between homotopy cocartesian squares of spectra. Since the three identity maps are obviously weak equivalences, the fourth map induced by the projection is a weak equivalence. But this map induces on homotopy groups the map $H_n^G(\mathrm{pr} :, E) : H_n^G(T :, E) \to H_n^G(\mathrm{pt})$ which is hence bijective for each $n \in \mathbb{Z}$.

This shows that $H_*^?(-; E)$ has the tree property if conditions (ii) and (iii) are satisfied. It is now also obvious that conditions (ii) and (iii) hold if $H_*^?(-; E)$ has the tree property.

emark 5.6

i. In the situation of Lemma 5.1(ii) diagram (5.3) is homotopy cocartesian if and only if the commutative diagram

$$
\begin{array}{ccc}
E(H) \vee E(H) & \xrightarrow{E(i_1) \vee E(i_2)} & E(G_1) \vee E(G_2) \\
{\scriptstyle \mathrm{id} \vee \mathrm{id}} \downarrow & & \downarrow {\scriptstyle E(j_1) \vee E(j_2)} \\
E(H) & \xrightarrow[E(j_0)]{} & E(G)
\end{array}
$$

where $j_0 : H \to G$ is defined to be $j_1 \circ i_1 = j_2 \circ i_2$, is homotopy cocartesian since there is a canonical weak equivalences from each corner of this square to the corresponding corner of (5.3).

ii. The situation in Lemma 5.1(iii) is a bit more complicated. The natural diagram to consider is

$$
\begin{array}{ccc}
E(H) \vee E(H) & \xrightarrow{\ E(i_0) \vee E(i_1)\ } & E(K) \\
{\scriptstyle \mathrm{id} \vee \mathrm{id}} \downarrow & & \downarrow {\scriptstyle E(j)} \\
E(H) & \xrightarrow[\ E(j \circ i_0)\ ]{} & E(G)
\end{array}
\tag{5.7}
$$

However, (5.7) is not commutative, while (5.4) is commutative. There is a canonical weak equivalence from each corner of (5.4) to the corresponding corner of (5.7), but those maps do not make the square

$$
\begin{array}{ccc}
E(H) \vee E(H) & \xrightarrow{\ E(i_0) \vee E(i_1)\ } & E(K) \\
\downarrow & & \downarrow \\
E^G(G/H) \vee E^G(G/H) & \xrightarrow[\ E^G(\mathrm{pr}_0) \vee E^G(\beta)\ ]{} & E^G(G/K)
\end{array}
$$

commutative.

The failure of the commutativity of (5.7) stems from the fact that the underlying diagram of groups commutes only up to conjugation, i.e. $j \circ i_0 \neq j \circ i_1 = c_t \circ j \circ i_0$. It is a consequence of the definitions that $E(c_t)$ is weakly homotopic to $\mathrm{id}_{E(G)}$, but in general there is no preferred homotopy. On the other hand, $E : \mathrm{GROUPOIDS} \to \mathrm{SPECTRA}$ is often slightly better than required in the discussion before Lemma 5.1, namely E is a 2-functor. This means that if τ is a natural transformation between functors f, g between groupoids, then there is a (preferred) homotopy $E(\tau)$ from $E(f)$ to $E(g)$. Under this stronger assumption on E there is a canonical homotopy that makes (5.7) homotopy commutative and then condition (iii) in Lemma 5.1 is equivalent to requiring that (5.7) is homotopy cocartesian with respect to the canonical homotopy.

ISOMORPHISM CONJECTURES FOR SPECTRA

In this section, we relate the (Fibered) Isomorphism Conjecture 1.1 for an equivariant homology theory $H_*^?$ to the versions appearing in Farrell–Jones [12] for algebraic K- and L-theory.

Consider a group homomorphism $\varphi: K \to G$, a K-CW-complex Z and a covariant functor E:SPACES→SPECTRA, which sends weak equivalences to weak equivalences and is compatible with disjoint unions , i.e. for a family $\{Y_i | i \in I\}$ of spaces the map induced by the inclusions $j_i: Y_i \to \coprod_{i \in I} Y_i$

$$\bigvee_{i \in I} \mathbf{E}(j_i): \bigvee_{i \in I} \mathbf{E}(Y_i) \to \mathbf{E}\left(\coprod_{i \in I} Y_i\right)$$

is a weak equivalence. We obtain a covariant functor

$$\mathbf{E}_Z^K: \mathrm{Or}(K) \to \mathrm{SPECTRA}, \quad K/H \mapsto \mathbf{E}(Z \times_K K/H).$$

Recall that for each covariant functor **F**: Or(K)→SPECTRA there is a K -homology theory $H_*^K(-:,\mathbf{F})$ defined for K-CW -complexes with the property that $H_n^K(K/H:\mathbf{F}) \cong \pi_n(\mathbf{F}(K/H))$ $H_n^K(K/H:\mathbf{F}) \cong \pi_n(\mathbf{F}(K/H))$ holds for $H \subseteq K$ and $n \in \mathbb{Z}$ [10, Sections 4, 7]. We denote by φ_*Z the G -space $G \times_\varphi Z$ obtained by induction with φ from the K-space Z. For a G-space X let φ^*X be the K-space obtained by restricting the G-action to a K -action using φ.

Lemma 6.1

For any G-CW-complex X there is an isomorphism, natural in X, Z and E,

$$\phi_*: H_n^K(\phi^*X; \mathbf{E}_Z^K) \xrightarrow{\cong} H_n^G(X; \mathbf{E}_{\phi_*Z}^G).$$

Proof

Let $\phi: \mathrm{Or}(K) \to \mathrm{Or}(G), K/H \to G/\phi(H)$ be the functor induced by φ. Given a contravariant (pointed) Or (G)-space A and a covariant (pointed) Or(K)-space B there is an adjunction

$$\mathrm{res}_\phi \, A \otimes_{\mathrm{Or}(K)} B \xrightarrow{\cong} A \otimes_{\mathrm{Or}(G)} \mathrm{ind}_\phi \, B, \tag{6.2}$$

where res_φ is restriction and ind_φ denotes induction with the functor $\varphi:\mathrm{Or}(K) \to \mathrm{Or}(G)$ (see [10, Lemma 1.9]). It induces a natural isomorphism

$$H_n^K (\phi^* X; \mathbf{E}_Z^K) \xrightarrow{\cong} H_n^G (X; \mathrm{ind}_\phi \mathbf{E}_Z^K).$$

There is a weak equivalence of covariant Or(G)-spectra

$$\mathrm{ind}_\phi \, \mathbf{E}_Z^K \to \cong \mathbf{E}_{\phi_* Z}^G$$

coming from

$$\mathrm{map}_G(G/\varphi(?), G/??)) \otimes_{\mathrm{Or}(K)} (Z \times_K K/?) \to \cong Z \times_\varphi G/??,$$

$$(f,(z,k?)) \mapsto (z, f(\varphi(k)\varphi(?)))$$

and the fact that **E** is compatible with disjoint unions.

Lemma 6.3

Let Fbe a family of subgroups of G. Let $N \in Z$. Then the following assertions are equivalent:

i. For any free G-CW-complex Z and $n \in Z$ (with $n \leq N$) the assembly map

$$H_n^G (E_{\mathcal{F}}(G); \mathbf{E}_Z^G) \xrightarrow{\cong} H_n^G (\mathrm{pt}; \mathbf{E}_Z^G)$$

is bijective.

ii. For each injective group homomorphism $\varphi: K \to G$ and any free connected K-CW-complex Z and $n \in Z$ (with $n \leq N$) the assembly map

$$H_n^K(E_{\phi^*(\mathscr{F})}(K); \mathbf{E}_Z^K) \overset{\cong}{\to} H_n^K(\text{pt}; \mathbf{E}_Z^K)$$

is bijective.

iii. For each group homomorphism $\varphi: K \to G$ and any free simply connected K-CW-complex Z and $n \in Z$ (with $n \leq N$) the assembly map

$$H_n^K(E_{\phi^*(\mathscr{F})}(K); \mathbf{E}_Z^K) \overset{\cong}{\to} H_n^K(\text{pt}; \mathbf{E}_Z^K)$$

is bijective.

Proof

(i) $\Rightarrow$ (ii) and (i) $\Rightarrow$ (iii): These implications follow from Lemma 6.1 since for any group homomorphism $\varphi: K \to G$ we have $\varphi^*(E_F(G)) = E_{\varphi^*(F)}(K)$.

(ii) $\Rightarrow$ (i): We can write a G-CW-complex Z as $Z = \coprod_{i \in I} G \times_{G_i} Z_i$ for subgroups $G_i \subseteq G$ and connected free G_i-CW-complexes Z_i. Since $\mathbf{E}$ is compatible with disjoint unions, we conclude from [10, Lemma 4.6] that we can assume, without loss of generality, that I consists of one element 0, i.e. $Z = G \times_{G_0} Z_0$. Now, the claim follows from Lemma 6.1 applied to the inclusion $\varphi: G_0 \to G$ and the free connected G_0-CW-complex Z_0.

(iii) $\Rightarrow$ (ii): There is an extension of groups $1 \to \pi 1(Z) \to \tilde{K} \to pK \to 1$ and a $\tilde{K}$-action on the universal covering $\tilde{Z}$ which extends the $\pi_1(Z)$-action on $\tilde{Z}$ and covers the K-action on Z. Moreover, $\tilde{Z}$ inherits the structure of a free $\tilde{K}$-CW-complex. Now, the claim follows from Lemma 6.1 applied to the group homomorphism $p: \tilde{K} \to K$ and the simply connected free $\tilde{K}$-CW-complex $\tilde{Z}$ since $p_*\tilde{Z} = Z$ and $p^*(\varphi^*F) = (\varphi \circ p)^*F$.

Lemma 6.4

Suppose that for any two-connected map f: X→Y the induced map E(f):E(X)→E(Y) is a weak equivalence. Let Z be a simply-connected G-CW-complex and f: Z→EG be the classifying map.

Then it induces a weak equivalence of Or (G)-spectra

$$f: \mathbf{E}_Z^G \to \mathbf{E}_{EG}^G$$

and, in particular, for each G-CW-complex and each n∈Z a natural isomorphism

$$H_n^G(X; \mathbf{E}_Z^G) \to \; \cong H_n^G(X; \mathbf{E}_{EG}^G).$$

Proof

The map f: Z→EG is 2-connected. Hence, the induced map $f_{\times_G} \mathrm{id}_{G/}$ ${}_H : Z_{\times_G} G/H \to EG_{\times_G} G/H$ is 2-connected for all subgroups H⊆G. Now apply [10, Lemma 4.6].

Definition 6.5 Fibered Isomorphism Conjecture for Spectra

We say that a group G satisfies the Isomorphism Conjecture for F and E (in the range ≤ N) if the assembly map induced by the projection pr:$E_F(G)$→pt

$$\mathrm{asmb}: H_n^G(E_{\mathscr{F}}(G); \mathbf{E}_{EG}^G) \to H_n^G(\mathrm{pt}; \mathbf{E}_{EG}^G)$$

is bijective for all n∈Z (with n≤N).

We say that a group G satisfies the Fibered Isomorphism Conjecture for F and E (in the range ≤N) if for any free G-CW-complex Z the assembly map induced by the projection pr:$E_F(G)$→pt

$$\mathrm{asmb}: H_n^G(E_{\mathscr{F}}(G); \mathbf{E}_Z^G) \to H_n^G(\mathrm{pt}; \mathbf{E}_Z^G)$$

is bijective for all n∈Z (with n ≤N).

Remark 6.6

The (Fibered) Isomorphism Conjecture of Farrell–Jones [12] for algebraic K-theory or L -theory, respectively, is equivalent to the (Fibered)
Isomorphisms Conjecture 6.5 for (G,VCYC,**E**) if **E**:SPACES→SPECTRA
sends X to the non-connective algebraic K -theory spectrum or L $^{\langle-\infty}$
$^{\rangle}$ -theory spectrum of the fundamental groupoid of X , respectively.
Since a 2-connected map f:X→Y induces an equivalence on the fundamental groupoids, Lemma 6.4 applies. Let R be a ring. Consider the
covariant functors

K_R: GROUPOIDS→SPECTRA,

L_R: GROUPOIDS→SPECTRA

defined in [10, Section 2] satisfying $\pi_n(K_R(G))=K_n(RG)$ and
$\pi_n(L_R(G))=L_n^{\langle-\infty\rangle}(RG)$ for each group G and n∈Z. Let $H_*^?(-,K_R)$ and
$H_*^?(-,L_R)$ be the associated equivariant homology theories. Then the
(Fibered) Isomorphism Conjecture for algebraic K-theory or algebraic
L-theory, respectively, for the group G in the sense of Farrell–Jones
[12] is equivalent to the (Fibered) Isomorphisms Conjecture 1.1 for
$H_*^?(-,K_R)$ and $H_*^?(-,L_R)$ for the pair (G, VCYC). This follows from Lemmas 6.3 and 6.4.

For more information about the various conjectures such as the version for pseudoisotropy or the Baum–Connes Conjecture we refer for
instance to [18].

THE *KH*-ISOMORPHISM CONJECTURE

In this section, we will formulate the KH-Isomorphism Conjecture.
The construction of homotopy algebraicK -Theory is a simplicial construction, so we will quickly fix the notation. The category Δ has as
objects finite ordered sets of the form n = {0 < 1 < ... < n} and order preserving maps as morphisms. The n -simplex $\Delta_{\bullet}^n$ is the simplicial set

$\underline{m} \mapsto \text{Map}_\Delta(\underline{m},\underline{n})$. Let R be a ring. The simplicial ring R[•] is defined by

$$R[\underline{n}] = R[t_0, \ldots, t_n]/(t_0 + \cdots + t_n = 1).$$

Here, the structure maps acts as follows: if $f : \underline{n} \to \underline{m}$ is order preserving then $f^* : R[\underline{m}] \to R[\underline{n}]$ is defined by

$$f^*(t_k) = \sum_{j \in f^{-1}(k)} t_j.$$

In [31] the homotopy algebraic K -theory $KH_*(R)$ of R is defined as the homotopy groups of the realization KHR of the simplicial spectrum $K^{-\infty}R[•]$. Here, $K^{-\infty}$ denotes the non-connected K-theory spectrum; a construction is reviewed before Definition 9.4. To illustrate the construction of homotopy algebraic K-theory we give a proof of the following fundamental property of homotopy algebraic K-theory, cf. [31, 1.2.(i)].

Proposition 7.1
The inclusion R↪R[X]gives an isomorphism $KH_n(R) \cong KH_n(R[X])$ for all n∈Z.

Proof
It suffices to show that R[•]→R[X][•] is a homotopy equivalence of simplicial rings (cf. Remark 7.2). To see this, we need to show that X↦0 is homotopic to the identity of R[X][•]. Such a homotopy R[X][•]×$\Delta^1_\bullet$ → R[X][•] is given by

$$(X, f) \mapsto \left(\sum_{j \in f^{-1}(0)} t_j \right) X.$$

where $f : \underline{n} \mapsto \underline{1}$.

Remark 7.2

If S is a ring and F is a finite set, then $S \times F$ has a ring structure $(S \times F \cong \prod_{f \in F} S)$. Therefore, we may view $R[X][\bullet] \times \Delta^1_{\bullet}$ as a simplicial ring and the above homotopy as a map of simplicial rings. Therefore, we get a map $\left| K^{-\infty}(R[X][\bullet] \times \Delta^1_{\bullet}) \right| \to \left| K^{-\infty}(R[X][\bullet]) \right|$. On the other hand, there is a map of simplicial spectra $K^{-\infty}(R[X][\bullet]) \times \Delta^1_{\bullet} \to K^{-\infty}(R[X][\bullet] \times \Delta^1_{\bullet})$ defined as follows. For $f \in \Delta^1_n$ there is an obvious map of rings $\quad$ If $: R[X][\underline{n}] \to R[X][\underline{n}] \times \{f\}$ $R[X][\underline{n}] \times \Delta^1_n$. Thus, we can map $(x, f) \in K^{-\infty}(R[X][\underline{n}]) \times \Delta^1_n$ to $K^{-\infty}(If)(x) \in K^{-\infty}(R[X][\underline{n}]) \times \Delta^1_n$.

In order to define an equivariant homology theory we define the functor

KH_R: GROUPOIDS→SPECTRA

as the realization of the simplicial functor

$K_{R[\bullet]}$:GROUPOIDS→SPECTRA.

Thus, $\pi_n(KH_R(G))=KH_n(RG)$. Since the realization of a weak equivalence is again a weak equivalence, KH_R sends equivalences of groupoids to weak equivalences of spectra.

Conjecture 7.3 (Fibered) *KH*-Isomorphism Conjecture

A group G is said to satisfy the (Fibered) KH-Isomorphism Conjecture (for a ring R) if the pair (G, FIN) satisfies the (Fibered) Isomorphism Conjecture 1.1 for the equivariant homology theory $H^?_*(-;KH_R)$.

Remark 7.4

All virtually cyclic groups act on trees with finite stabilizers. For a finite group F the (Fibered) KH-Isomorphism Conjecture holds (since $EF_{IN}(F)=F/F$). Thus, by Theorem 0.5 the (Fibered) KH-Isomorphism Conjecture holds for virtually cyclic groups. Therefore, Theorem 2.4 implies that it makes no difference if we replace the family of finite groups with the family of virtually cyclic groups in the formulation of the (Fibered) KH-Isomorphism Conjecture.

THE RELATION BETWEEN THE K - AND THE KH-ISOMORPHISM CONJECTURE

There is a natural map $K^{-\infty}R \to KHR$ induced from the inclusion of the constant simplicial ring R into $R[\bullet]$. Similarly, we obtain a natural transformation $K_R \to KH_R$ of functors from GROUPOIDS to SPECTRA. Thus, we obtain a natural transformation of equivariant homology theories

$H_*^?(-;K_R) \to H_*^?(-;KH_R)$ and a commutative diagram between assembly maps

$$H_n^G(E_{\mathcal{FIN}}(G); \mathbf{K}_R) \longrightarrow K_n(RG)$$

$$H_n^G(E_{\mathcal{FIN}}(G); \mathbf{KH}_R) \longrightarrow KH_n(RG) \qquad\qquad (8.1)$$

We first explain what the KH-Isomorphism Conjecture 7.3 implies for the K-Isomorphism Conjecture, i.e. the Farrell–Jones Conjecture for algebraic K-theory (see Remark 6.6). In order to state the connection, we need to recall the groups $N^p K_n(R)$ [6, XIII]. They can be defined by $N^0 K_n(R) = K_n(R)$ and

$$N^p K_n(R) = ker(N^{p-1}(q): N^{p-1} K_n(R[t]) \to N^{p-1} K_n(R)),$$

where $q(t) = 0$. For regular rings $N^p K_n(R) = 0$ for $p \geq 1$, see [13].

Proposition 8.2
Let G be a group that satisfies the KH-Isomorphism Conjecture 7.3 for the ring R.

i.   Suppose that $N^p K_n(RF)=0$ for all finite subgroups F of G and all $n \in Z$, $p \geq 1$. Then the assembly map with respect to the family FIN in algebraic K-theory, i.e. the top row in (8.1), is split injective.

ii. Suppose that $N^p K_n(RF) \otimes Q=0$ for all finite subgroups F of G and all $n \in Z$, $p \geq 1$. Then assembly map with respect to the family FIN in algebraic K-theory, i.e. the top row in (8.1), is rationally split injective.

Proof

By the spectral sequence from [31, 1.3] the canonical map $K_*(A) \to KH_*(A)$ is an isomorphism if $N^p K_n(A)=0$ for all $n \in Z$ and $p \geq 1$ and a rational isomorphism if $N^p K_n(A) \otimes Q=0$ for all $n \in Z$ and $p \geq 1$. Therefore, these assumptions imply by a spectral sequence argument that the left vertical map in (8.1) is an isomorphism or a rational isomorphism, respectively.

Remark 8.3

The assumptions of Proposition 8.2(i) and (ii) are satisfied in the following cases:

I.   If R is a regular ring containing Q then RF is regular for all finite groups F. Thus, the assumption in 8.2(i) is satisfied.

II. If R=Z then the assumption in 8.2(ii) is satisfied. This follows from [30, 6.4], which implies

$$NK_*(Z[t_1,\dots,t_n]F) \otimes Q \cong NK_*(Q[t_1,\dots,t_n]F).$$

Thus, for a finite group F it follows that $NK_*(Z[t_1,\dots,t_n]F)$ vanishes rationally, since $Q[t_1,\dots,t_n]F$ is regular. A straightforward induction shows that this implies $N^p K_n(ZF) \otimes Q=0$ for $p \geq 1$.

III. If G is torsionfree and R is regular, then the assumption in 8.2(i) is satisfied. In this case, the Farrell–Jones Conjecture in algebraic K-theory asserts, that the top vertical map in (8.1) is an isomorphism. Thus, in this situation (R regular, G torsionfree) the Farrell–Jones

Conjecture in algebraic K-theory holds if $N_n^p(RG)=0$ for all $n \in Z$ and $p \geq 1$ and if G satisfies the KH-Isomorphism Conjecture.

Next, we explain the reverse connection.

Theorem 8.4 The K -theory version implies the KH-version

Let G be a group and let R be a ring.

(i)Suppose that the (Fibered) Farrell–Jones Conjecture in algebraic K-theory is true for $(G, R[x_1, x_2, \ldots, x_n])$ for all $n \leq 1$ then the (Fibered) KH-Isomorphism Conjecture 7.3 is true for (G, R). (ii)Suppose that the Fibered Farrell–Jones Conjecture in algebraic K-theory is true for $(G \times \mathbb{Z}^n, R)$ for all $n \leq 1$. Then the Fibered KH-Isomorphism Conjecture 7.3 is true for (G, R).

Proof

Let $\varphi : K \to G$ be a group homomorphism. The assembly map $H_n^K(E_{\phi^*}$
$\mathcal{V}\mathcal{C}\mathcal{Y}\mathcal{C}(K); K_{R[\underline{n}]}) \to K_n(R[\underline{n}][K])$ is on the level of spectra given by the map

$$E_{\phi^* \mathcal{V}\mathcal{C}\mathcal{Y}\mathcal{C}}(K) \otimes_{Or(K)} K_{R[\underline{n}]} \to K/K \otimes_{OrK} K_{R[\underline{n}]} \simeq K_{R[\underline{n}]}(K)$$

induced by $E_{\varphi^* VCYC}(K) \to K/K$. The assumption in (i) is that this map of spectra is a weak equivalence. Using the fact that the realization of a map of simplicial spectra that is levelwise a weak equivalences is a weak equivalence and the identification

$$|E_{\varphi^* VCYC}(K) \otimes_{Or(K)} K_{R[\bullet]}| \cong E_{VCYC}(K) \otimes_{Or(K)} |K_{R[\bullet]}|,$$

we conclude that the (Fibered) Farrell–Jones Conjecture for $(G, R[x_1, \ldots, x_n])$ for all n implies the (Fibered) KH-Isomorphism Conjecture for (G, R) with the family of finite subgroups replaced by the family of virtually cyclic subgroups. By Remark 7.4 this is equivalent to the (Fibered) KH-IsomorphismConjecture 7.3 for (G, R).

Next, we prove (ii) by reducing it to (i). For a group K we denote by $p_K : K \times \mathbb{Z} \to K$ the canonical projection. We observe first that the (Fibered) Isomorphism Conjecture for $(G, VCYC, R[\mathbb{Z}])$ is equivalent to the (Fibered) Isomorphism Conjecture $(G \times \mathbb{Z}, (p_G)^* VCYC, R)$ because for every group K and every K-spaceX there is a natural isomorphism

$$H_n^K(X; \mathbf{K}_{R[\mathbb{Z}]}) \cong H_n^{K \times \mathbb{Z}}(p_K^* X; \mathbf{K}_R),$$

where $p_K^* X$ denotes the $K \times Z$-space obtained by restriction of X along p_K and because $p_K^* E_F(K) = E_{(PK)^*F}(K \times \mathbb{Z})$ for every family of subgroups of K. If the Fibered Isomorphism Conjecture holds for a family F, then it will also hold for every family G that contains F [4, Lemma 1.6]. Because the family of virtually cyclic subgroups of G×Z is contained in $(p_G)^*$VCYC the Fibered Farrell–Jones Conjecture for (G×Z,R) implies the Fibered Farrell–Jones Conjecture for (G,R[Z]). By the Bass–Heller–Swan splittings [14], the latter is equivalent to the Fibered Farrell–Jones Conjecture for(G,R[x]). By induction on n this means that the assumption in (ii) implies the assumption of (i).

Remark 8.5
It is not unreasonable to expect that the non-fibered version of Theorem 8.4(ii) is also valid. Our argument would also prove the non-fibered version if we were to know that for every virtually cyclic group V the product V×Z satisfies the Farrell–Jones Conjecture. This seems very likely, but we could not find such a statement in the literature.

Remark 8.6
Let us briefly list some consequences of the Farrell–Jones Conjecture for algebraic K-theory. Suppose that the Farrell–Jones Conjecture for algebraic K-theory holds for the group G and every regular ring R. Now, consider a group G and a regular ring R   with the property that either Q⊆R holds or G is torsionfree. The proof of Proposition 8.2, Remark 8.3 and Theorem 8.4 imply:

I.   The KH-Isomorphism Conjecture 7.3 is true for G and R.
II. The canonical map $K_n(RG) \to KH_n(RG)$ is bijective for n∈Z.
III. $N^p K_n(RG)=0$ for p⩾1 and n∈Z, see [3, Proposition 7.4].

Remark 8.7 Injectivity of the KH-Assembly Map

In many cases injectivity of the assembly map

$$H_n^G(E_{\mathcal{F}\mathcal{I}\mathcal{N}}(G); \mathbf{K}_R) \to K_n(RG)$$

is proven by construction of a spectrum $\mathbf{T}(R,G)$ and a map of spectra $K^{-\infty}RG \to T(R,G)$ such that for many groups the composition of the assembly map on the level of spectra with this map is a weak equivalence. The construction of $K^{-\infty}RG \to T(R,G)$ is always natural in the coefficient ring R. Therefore, applying the arguments of the proof of Theorem 8.4(i) we can use $T(R[\bullet],G)$ to split theKH-assembly map in this cases. This proves that the KH-assembly map is split injective for groups G of finite asymptotic dimension that admit a finite model for BG [2] and for groups G for which $E_{FIN}(G)$has a compactification with certain properties [23].

NON-CONNECTIVE WALDHAUSEN NIL

Before we can show that $H_*^?(-;KH_R)$ has the tree property, we will need to recall Waldhausen's work on K-theory of generalized free products [29]. We start with Waldhausen's Nil-groups.

Definition 9.1 Nil-Categories

Let R be a ring and X,Y,Z,W be R-bimodules.

(i)The category NIL(R;X,Y) has as objects quadruples (P,Q,p,q), where P and Q are finitely generated projective R -modules and $p:P \to Q \otimes_R X$, $q:Q \to P \otimes_R Y$ are R-linear maps subject to the following nilpotence condition: let $P_0 = Q_0 = 0$, $P_{n+1} = p^{-1}(Q_n \otimes_R X)$ and $Q_{n+1} = q^{-1}(P_n \otimes_R Y)$. It is required that for sufficient large N, $P = P_N$ and $Q = Q_N$.

(ii) The category $NIL(R;X,Y,Z,W)$ has as objects quadruples (P,Q,p,q), where P and Q are finitely generated projective R-modules and $p:P\to Q\otimes_R X\oplus P\otimes_R Z$, $q:Q\to P\otimes_R Y\oplus Q\otimes_R W$ are R-linear maps subject to the following nilpotence condition: let $P_0=Q_0=0$, $P_{n+1}=p^{-1}(Q_n\otimes_R X\oplus P_n\otimes_R Z)$ and $Q_{n+1}=q^{-1}(P_n\otimes_R Y\oplus Q_n\otimes_R W)$. It is required that for sufficient large N, $P=P_N$ and $Q=Q_N$.

Morphisms are in both cases R-linear maps $P\to P'$, $Q\to Q'$ that are compatible with p, p', q and q'. Both categories are exact categories, where sequences are exact whenever they map to exact sequences of modules under $(P,Q,p,q)\mapsto P$ and $(P,Q,p,q)\mapsto Q$.

Remark 9.2.

Let $f_R:R\to S$ be a map of rings and $f_X:X\to X'$, $f_Y:Y\to Y'$ be maps over f_R, i.e. X' and Y' are S-bimodules, $f_X(rxr')=f_R(r)f_X(x)f_R(r')$ and similar for f_Y. Then (f_R,f_X,f_Y) induce an exact functor $NIL(R;X,Y)\to NIL(S;X',Y')$ sending (P,Q,p,q) to $(P\otimes_R S,Q\otimes_R S,p_S,q_S)$, where p_S and q_S are the canonical maps. For example, p_S is the composition

$$P\otimes_R S\to Q\otimes_R X\otimes_R S\to Q\otimes_R X'\cong Q\otimes_R S\otimes_S X',$$

where the first map uses p and the second uses f_X and left multiplication of S. In particular, we get a functor $S\otimes-:NIL(R;X,Y)\to NIL(S\otimes R;S\otimes X,S\otimes Y)$. If $f:S\to S'$ is a map of rings, then we get another functor $f_*:NIL(S\otimes R;S\otimes X,S\otimes Y)\to NIL(S'\otimes R;S'\otimes X,S'\otimes Y)$. The functoriality of $NIL(R;X,Y,Z,W)$ is similar.

We review next [28, 2.5] in a slightly more modern language and discuss applications to Waldhausen Nil-categories.

A sum ring is a ring S together with elements v, $\bar{v}$, u and $\bar{u}$ of S such that $u\bar{u}=1$, $v\bar{v}=1$ and $\bar{v}v+\bar{u}u=1$. This implies that $u\bar{v}=0$ and $v\bar{u}=0$. Moreover, the map $f_\oplus:S\oplus S\to S$ defined by $(r,s)\mapsto \bar{u}ru+\bar{v}sv$ is a ring homomorphism. Let M,N be S-modules. Denote by (M,N) the direct sum $M\oplus N$ considered as an $S\oplus S$-module. The S-modules $(M,N)\otimes f_\oplus S$

and M⊕N (considered as anS-module as usual) are naturally isomorphic. Such an isomorphism and its inverse are given by

$$M \oplus N \ni m \oplus n \mapsto (m, n) \otimes (\bar{u} + \bar{v}) \in (M, N) \otimes_{f_\oplus} S,$$

(M,N)⊗f$_\oplus$S∋(m,n)⊗r↦mur⊕nvr∈M⊕N.

An infinite sum ring is a sum ring together with a ring endomorphism
f$_\infty$ such that f$_\oplus$(r,f$_\infty$(r))=f$_\infty$(r).

Remark 9.3

The functor M↦M⊗f$_\infty$S is an Eilenberg swindle on the category P$_S$ of
finitely generated projective modules over such an infinite sum ring.
Indeed,

M⊗f$_\infty$S=M⊗f$_{\oplus\circ}$(id$_s$,f$_\infty$)S≅M⊗$_(id$_s$,f$_\infty$)(S⊕S)⊗$_{f\oplus}$S≅(M,M⊗$_{f\infty}$S)⊗$_{f\oplus}$S≅M⊕(
M⊗$_{f\infty}$S).

The same swindle applies to Waldhausen's Nil-categories: fix an infinite sum ring S. Let X and Y be bimodules over another ring R. Then
the endofunctor (f$_\infty$)$_*$ is equal to the composition

$$\text{NIL}(S \otimes R; S \otimes X, S \otimes Y)$$

$$(\text{id}, f_\infty)_* \Big\downarrow$$

$$\text{NIL}((S \oplus S) \otimes R; (S \oplus S) \otimes X, (S \oplus S) \otimes Y)$$

$$(f_\oplus)_* \Big\downarrow$$

$$\text{NIL}(S \otimes R; S \otimes X, S \otimes Y).$$

Using the natural isomorphism from above there is a natural transformation from this composition toid ⊕ (f$_\infty$)$_*$. Thus, (f$_\infty$)$_*$ is an Eilenberg swindle. This swindle is compatible with the two forgetful func-

tors $NIL(S \otimes R; S \otimes X, S \otimes Y) \to P_s \otimes_R$. Analogous considerations apply to $NIL(S \otimes R; S \otimes X, S \otimes Y, S \otimes Z, S \otimes Z, S \otimes W)$.

The cone ring ΛZ of Z is the ring of column and row finite $N \times N$-matrices over Z, i.e. matrices such that every column and every row contains only finitely many non-zero entries. The suspension ring ΣZ is the quotient of ΛZ by the ideal of finite matrices. For an arbitrary ring R we define $\Lambda R = \Lambda Z \otimes R$ and $\Sigma R = \Sigma Z \otimes R$. We will view Λ and Σ as functors. Every bijection $N \to N \times N$ induces the structure of an infinite sum ring on the cone ring ΛR, cf. [28, p. 355]. We can consider ΛR as a subring of the ring of column and row finite $N \times N$-matrices over R. However, this inclusion is not always an equality.

Next, we want to define a non-connective spectrum associated to Waldhausen's Nil-categories. First, we recall the construction for K -theory. Denote by KR the K -theory space of a ring (obtained, for example, by applying Waldhausen's $_s.$-construction to the category $_{PR}$ of finitely generated projective R-modules). Then th space of the spectrum $K^{-\infty}R$ is by definition $K\Sigma^nR$. The composition $KR \to K\Lambda R \to K\Sigma R$ is constant. The choice of an bijection $N \to N \times N$ gives an Eilenberg swindle on ΛR, cf. Remark 9.3. If we fix such a bijection we get a functorial way of contracting $K\Lambda R$ to the basepoint. This induces the structure maps $\Sigma(K\Sigma^nR) \to K\Sigma^{n+1}R$.

Definition 9.4 Non-Connective Nil-Spectra

Let R be a ring and X and Y be R -bimodules. The (non-connective) spectrum $NiL^{-\infty}(R;X,Y)$ has $KNIL(\Sigma^nR; \Sigma^nX, \Sigma^nY)$ as its n th space. Here, $\Sigma^nX = \Sigma^nZ \otimes X$ and we define similarly Σ^nY, ΛX and ΛY. The structure maps are defined in an analogous ways as for the non-connective K-theory spectrum: the functoriality discussed in Remark 9.2 allows us to consider the (constant) composition

$NIL(C;A',B') \to NIL(\Lambda C; \Lambda A', \Lambda B') \to NIL(\Sigma C; \Sigma A', \Sigma B')$.

The structure maps for $\mathrm{NiL}^{-\infty}(R;X,Y)$ are now defined using the Eilenberg swindle on the second category discussed in Remark 9.3. Similarly, we define a (non-connective) spectrum $\mathrm{NiL}^{-\infty}(R;X,Y,Z,W)$ with

$$\mathrm{KNIL}(\Sigma^n R;\Sigma^n X,\Sigma^n Y,\Sigma^n Z,\Sigma^n W)$$

as its nth space.

An inclusion $\alpha:C\to A$ of rings is called pure if $A=\alpha(C)\oplus A'$ as C-bimodules. It is called pure and free if in addition A' is free as a left C -module. If $H\to G$ is an inclusion of groups, then the inclusion $RH\to RG$ of rings is pure and free. The following observation is straightforward.

Lemma 9.5
If α is pure (and free) then $\Sigma\alpha$ and $\Lambda\alpha$ are also pure (and free).

Let $\alpha: C\to A$ and $\beta: C\to B$ be both pure. The ring $R=A*_C B$, the free product of A and B, amalgamated at C (w.r.t. α, β), is defined by the pushout

$$
\begin{array}{ccc}
C & \xrightarrow{\alpha} & A \\
{\scriptstyle\beta}\downarrow & & \downarrow \\
B & \longrightarrow & R.
\end{array}
$$

For group rings this corresponds to amalgamated products of groups.

Lemma 9.6
The cone, respectively suspension ring, of $A*_C B$ is naturally isomorphic to $\Lambda A*_{\Lambda C}\Lambda B$, respectively $\Sigma A*_{\Sigma C}\Sigma B$.

Proof
This follows from the universal property. \

Let α, β: C→A be pure and free. The Laurent extension w.r.t. α and β is the universal ring R=A$_{\alpha,\beta}${t$^{\pm 1}$}that contains A and an invertible element t and satisfies

$$\alpha(c)t = t\beta(c) \quad \text{for } c \in C.$$

Existence is explained in [29, p. 149]. For group rings this corresponds to HNN extensions.

Lemma 9.7

The cone, respectively suspension ring, of $A_{\alpha,\beta}${t$^{\pm 1}$}is naturally isomorphic to $\Lambda_{A\Lambda\alpha,\Lambda\beta}${T$^{\pm 1}$},respectively $\Sigma_{A\Sigma\alpha,\Sigma\beta}${T$^{\pm 1}$}.

Proof

This follows from the universal property.

WALDHAUSEN'S CARTESIAN SQUARES

Let α:C→A and β:C→B be pure and free. Write A=α(C)⊕A$'$ and B=β(C)⊕B$'$ as C -bimodules. LetR=A$^*_C$B. Consider the square

$$
\begin{array}{ccc}
NIL(C; A', B') & \longrightarrow & \mathscr{P}_A \times \mathscr{P}_B \\
\downarrow & & \downarrow \\
\mathscr{P}_C & \longrightarrow & \mathscr{P}^*_R
\end{array}
\tag{10.1}
$$

The two functors starting at the upper left-hand corner are defined by sending (P,Q,p,q) to(P⊕Q)$\in_{PC}$, respectively to (P$_{\otimes\alpha}$A,Q$_{\otimes\beta}$B). The category P$^*_R$ is defined in [29, p. 205]. It is a cofinal full subcategory of P$_R$ and contains all finitely generated free modules. There is an obvious natural transformation between the two ways to go through the diagram. However, there is also a not quite so obvious more complicated natural transformation that makes use of p and q, cf. [29, 1.4, 11.3]: let$_{iP}$:$P\otimes_C A\otimes_A R\to P\otimes_C R$ and iQ:$Q\otimes_C B\otimes_B R\to Q\otimes_C R$ be the natural isomorphisms. Define N by the commutative diagram

$$
\begin{array}{ccc}
P & \xrightarrow{\;\;p\;\;} & Q\otimes_C A \\
\downarrow & & \downarrow \\
P\otimes_C A\otimes_A R & \xrightarrow{\;\;N\;\;} & Q\otimes_C R
\end{array}
$$

and similarly M:$Q\otimes_C B\otimes_B R\to P\otimes_C R$. Then

$$
\begin{pmatrix} i_P & M \\ N & i_Q \end{pmatrix}
$$

is an isomorphism and defines the more complicated natural transformation. It is a result of Waldhausen[29, 11.3], that applying K to (10.1) yields a homotopy cartesian square (w.r.t. the homotopy induced by the more complicated natural transformation). We will need a non-connective version of Waldhausen's result.

Theorem 10.2 Non-connective Versions of Waldhausen's Homotopy Cartesian Squares for Amalgamation
We have the following diagram of spectra:

$$
\begin{array}{ccc}
\mathrm{NiL}^{-\infty}(C;\,A',\,B') & \longrightarrow & K^{-\infty}A\wedge K^{-\infty}B \\
\downarrow & & \downarrow \\
K^{-\infty}C & \longrightarrow & K^{-\infty}R
\end{array}
$$

The more complicated natural transformations combine to a homotopy between the two ways to go through this diagram. The diagram is homotopy cartesian w.r.t. this homotopy.

Proof

The diagram of spectra is obtained from (10.1) by tensoring everything in sight by $\Sigma^n Z$ (and applying K). We need to check compatibility with the structure maps. Those come from an intermediate diagram where we apply $\Lambda Z \otimes$- and use an Eilenberg swindle. This Eilenberg swindle happens on the left of this tensor product, while everything else happens on the right. This proves compatibility with the structure maps.

If we use P_R^* rather than P_R then the diagram is homotopy cartesian by Waldhausen's result and Lemma 9.6. However, since the former category contains all finitely generated free modules and we use non-connective K -theory we can also use P_R.

Remark 10.3 Waldhausen Nil for Amalgamations Vanishes for Regular Coherent Rings

Waldhausen proved that for a regular coherent ring C, the functor $P_C \times P_C \to \mathrm{NIL}(C; A',B')$ defined by $(P,Q) \mapsto (P,Q,0,0)$ induces an isomorphism in connective K-theory, [29, 12.2]. A priori, this does not immediately imply that the induced map $\alpha : K^{-\infty}C \vee K^{-\infty}C \to \mathrm{NiL}^{-\infty}(C; A',B')$ is a weak equivalence because it is not clear whether ΣC is again regular coherent. However, if C is regular or more generally, if C is a group ring with a regular coefficient ring over a regular coherent group, then α is a weak equivalence. This can be seen as follows. The functor $(P,Q,p,q) \mapsto (P,Q)$ splits α, thus α will be injective on homotopy groups. To prove surjectivity we use the fact that there is a C natural map of rings $C[Z^n] \to \Sigma^n C$, that is naturally split surjective in connective K-theory [28, Section 6]. We get the following commutative diagram:

$$
\begin{array}{ccc}
\mathscr{P}_{C[Z^n]} \times \mathscr{P}_{C[Z^n]} & \longrightarrow & \mathrm{NIL}(C[Z^n]; A'[Z^n], B'[Z^n]) \\
\downarrow & & \downarrow \\
\mathscr{P}_{\Sigma^n C} \times \mathscr{P}_{\Sigma^n C} & \longrightarrow & \mathrm{NIL}(\Sigma^n C; \Sigma^n A', \Sigma^n B')
\end{array}
\qquad (10.4)
$$

Apply $C[Z^n] \to \Sigma^n C$ map to the long exact sequence obtained from (10.1) by [29, 11.3]. A little diagram chase in the resulting ladder diagram shows that the right vertical map in (10.4) is surjective in connective K-theory. The assumptions on C imply that $C[Z^n]$ is regular coherent. Therefore, the top horizontal map in (10.4) is an isomorphism in connective K-theory. Therefore, the bottom horizontal map in (10.4) is also surjective in connective K-theory. This implies that α is surjective on homotopy groups.

Next, we discuss the analogous cartesian square for Laurent extensions. Let $\alpha, \beta : C \to A$ be pure and free and $R =_{A\alpha, \beta} \{t^{\pm 1}\}$. We denote by $\iota : A \to R$ the inclusion. Write $A = \alpha(C) \oplus A'$ and $A = \beta(C) \oplus A''$ as C-bimodules. Consider the square

$$
\begin{array}{ccc}
NIL(C; {}_\alpha A'_{\alpha, \beta} A''_{\beta, \beta} A_{\alpha, \alpha} A_\beta) & \longrightarrow & \mathcal{P}_A \\
\downarrow & & \downarrow {\scriptstyle \iota_*} \\
\mathcal{P}_C & \xrightarrow[(\iota \circ \alpha)_*]{} & \mathcal{P}^*_R
\end{array}
\qquad (10.5)
$$

Here, we use α and β to indicate the C-bimodule structures. The two functors starting at the upper right-hand corner are defined by sending (P, Q, p, q) to $(P \oplus Q) \in_{PC}$, respectively to $(P \otimes_\alpha A \oplus Q \otimes_\beta A)$. The category P^*_R is defined in [29, p. 205]. It is a cofinal full subcategory of P_R and contains all finitely generated free modules. As before, there is an obvious and a more complicated natural transformation between the two ways to go through the diagram [29, 2.4, 12.3]: let $i_p : P \otimes_\alpha A \otimes_A \otimes R \to P \otimes_\alpha R$ and $i_Q : Q \otimes_\beta A \otimes R \to Q \otimes_\alpha R$ denote the canonical isomorphisms. (Here, i_Q uses an extra t, i.e. $i_Q(y \otimes a \otimes r) = y \otimes tar$.) These isomorphisms give the obvious natural transformation. The more complicated natural transformation is obtained by adding a nilpotent term which we review next. Write $p = p_0 + p_1$, where $p_0 : P \to P \otimes_\beta A_\alpha$ and $p1 : P \to Q \otimes_\alpha A'_\alpha$. Define N_0 and N_1 by the commutative diagrams

$$
\begin{array}{ccc}
P & \xrightarrow{p_0} & P\otimes_\beta A_\alpha \\
\downarrow & & \downarrow \\
P\otimes_\alpha A\otimes_A R & \xrightarrow{N_0} & P\otimes_\alpha R
\end{array}
\qquad
\begin{array}{ccc}
P & \xrightarrow{p_1} & Q\otimes_\beta A'_\alpha \\
\downarrow & & \downarrow \\
P\otimes_\alpha A\otimes_A R & \xrightarrow{N_1} & Q\otimes_\alpha R
\end{array}
$$

(The second vertical arrow is $x\otimes a\mapsto x\otimes ta$.) Write $q=q_0+q_1$, where $q_0: Q\to Q\otimes_\alpha A_\beta$ and $q1: Q\to P\otimes_\beta A''_\beta$. Define M_0 and M_1 by the commutative diagrams

$$
\begin{array}{ccc}
Q & \xrightarrow{q_0} & Q\otimes_\alpha A_\beta \\
\downarrow & & \downarrow \\
Q\otimes_\beta A\otimes_A R & \xrightarrow{M_0} & Q\otimes_\alpha R
\end{array}
\qquad
\begin{array}{ccc}
Q & \xrightarrow{q_1} & P\otimes_\alpha A''_\beta \\
\downarrow & & \downarrow \\
Q\otimes_\beta A\otimes_A R & \xrightarrow{M_1} & P\otimes_\alpha R
\end{array}
$$

(The fourth vertical arrow is $x\otimes a\mapsto x\otimes ta$.) The more complicated natural transformation is then given by the isomorphism

$$
\begin{pmatrix} i_P & 0 \\ 0 & i_Q \end{pmatrix} + \begin{pmatrix} N_0 & M_1 \\ N_1 & M_0 \end{pmatrix}.
$$

It is a result of Waldhausen [29, 12.3], that applying K to (10.5) yields a homotopy cartesian square (w.r.t. the homotopy induced by the more complicated natural transformation). The arguments used to prove Theorem 10.2 can be used to prove a non-connective version of this result.

Theorem 10.6 Non-connective Versions of Waldhausen's Homotopy Cartesian Squares for Laurent Extensions
We have the following diagram of spectra:

$$\text{NiL}^{-\infty}(C;{}_{\alpha}A'_{\alpha,\beta}A''_{\beta,\beta}A_{\alpha,\alpha}A_{\beta}) \longrightarrow K^{-\infty}(A)$$

$$\downarrow \qquad\qquad\qquad\qquad \downarrow K^{-\infty}(\iota)$$

$$K^{-\infty}(C) \xrightarrow[\;\;K(\iota\circ\alpha)\;\;]{} K^{-\infty}(R)$$

The more complicated natural transformations combine to a homotopy between the two ways to go through this diagram. The diagram is homotopy cartesian w.r.t. this homotopy.

Remark 10.7 Waldhausen Nil for Laurent Extensions Vanishes for Regular Coherent Rings.

The reasoning in Remark 10.3 also applies to $\text{Nil}(C;_{\alpha} A'_{\alpha,\beta}A_{\alpha,\alpha}A_{\beta})$. If C is a group ring with a regular coefficient ring over a regular coherent group, then the functor $(P,Q) \mapsto (P,Q,0,0)$ induces a weak equivalence.

$$K^{-\infty}C \vee K^{-\infty}C \to \text{NiL}^{-\infty}(C;_{\alpha} A'_{\alpha,\beta}A''_{\beta,\beta}A_{\alpha,\alpha}A_{\beta}).$$

THE TREE PROPERTY FOR HOMOTOPY *K*-THEORY

This section contains the proof of the following result.

Theorem 11.1 Continuity and Tree-Property for $H^?_*(-;KH_R)$

The equivariant homology theory $H^?_*(-;KH_R)$ is continuous and has the tree property.

Let X, Y, Z and W be bimodules over R. Consider the simplicial spectra $\underline{n} \mapsto \text{NiL}(\mathbb{Z}[\underline{n}] \otimes R; \mathbb{Z}[\underline{n}] \otimes X, \mathbb{Z}[\underline{n}] \otimes Y$ and $\underline{n} \mapsto \text{NiL}(\mathbb{Z}[\underline{n}] \otimes R; \mathbb{Z}[\underline{n}] \otimes X, \mathbb{Z}[\underline{n}] \otimes Y, \mathbb{Z}[\underline{n}] \otimes Z, \mathbb{Z}[\underline{n}] \otimes W)$. We will denote the realization of these simplicial spectra by NH(R; X, Y) or NH(R;X,Y,Z,W), respectively. However, the point here is that this process kills the additional information in Waldhausen's Nil-groups.

Proposition 11.2

i. The functor $P_R \times P_R \to NIL(R;X,Y)$ defined by $(P,Q) \mapsto (P,Q,0,0)$ induces an equivalence $KH(R) \vee KH(R) \to NH(R;X,Y)$.

ii. The functor $P_R \times P_R \to NIL(R;X,Y,Z,W)$ defined by $(P,Q) \mapsto (P,Q,0,0)$ induces an equivalence $KH(R) \vee KH(R) \to NH(R;X,Y,Z,W)$.

Proof

We prove only (i), the proof of (ii) is similar. It suffices to show that the functor $(P,Q,p,q) \mapsto (P,Q,0,0)$ mapping the simplicial category $N(\bullet) = NIL(Z[\bullet] \otimes R; Z[\bullet] \otimes X, Z[\bullet] \otimes Y)$ to itself is simplicially homotopic to the identity. Such a homotopy $N(\bullet) \times \Delta^1_\bullet \to N(\bullet)$ is given by

$$(P, Q, p, q) \mapsto \left(P, Q, \sum_{j \in f^{-1}(0)} t_j \otimes p, \sum_{j \in f^{-1}(0)} t_j \otimes q \right)$$

where $f : \underline{n} \mapsto \underline{1}$.

The homotopy algebraic K-theory of a free product or a Laurent extension does, therefore, not involve Nil-groups.

Theorem 11.3 Homotopy Cartesian Squares for Homotopy K-theory

i. Consider the free product $R = A *_C B$ (w.r.t. pure and free maps $\alpha : C \to A$ and $\beta : C \to B$). Then the commutative diagram

$$
\begin{array}{ccc}
KH(C) \vee KH(C) & \xrightarrow{KH(\alpha) \vee KH(\beta)} & KH(A) \vee KH(B) \\
{\scriptstyle id \vee id}\downarrow & & \downarrow{\scriptstyle KH(\iota_A) \vee KH(\iota_B)} \\
KH(C) & \xrightarrow[KH(\alpha \circ \iota_A)]{} & KH(R)
\end{array}
$$

is homotopy cartesian. Here, ι_A and ι_B are the obvious inclusions of rings.

ii. Consider the Laurent extension $R = A_{\alpha,\beta}\{t^{\pm1}\}$ (w.r.t. pure and free maps $\alpha : C \to A$ and $\beta : C \to A$). Then the diagram

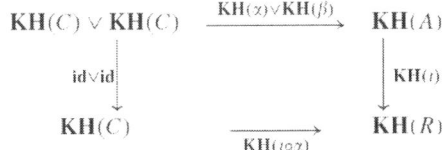

$$KH(C) \vee KH(C) \xrightarrow{\ KH(\alpha) \vee KH(\beta)\ } KH(A)$$

is homotopy cartesian w.r.t. the obvious homotopy between the two ways to go through this diagram, cf. Section 10. Here, i is the obvious inclusions of rings.

Proof

The realization of a simplicial diagram of spectra that is degreewise homotopy cartesian is again homotopy cartesian. Thus, the result follows by combining Theorems 10.2 or 10.6, respectively, and Proposition 11.2. There is no longer a difference between the obvious and the more complicated natural transformation, since we got rid of the nil-categories.

Proof (Proof of Theorem 11.1)

We use Lemma 5.1. We discuss first continuity, i.e. condition 5.1(i). For K_R this follows from the compatibility of K-theory with directed colimits [22]. Since realizations of simplicial spectra commutes with directed colimits, this implies condition 5.1(i).

Next, we discuss the tree property, i.e. conditions 5.1(ii) and (iii). Note that KH_R is a 2-functor as discussed in Remark 5.6. For K_R this holds since natural equivalences between functors F and G induce naturally a homotopy from $K^{-\infty}(F)$ to $K^{-\infty}(G)$. This homotopy is preserved under realization. Now, observe that the obvious homotopy in Theorem 11.3(ii) is in the case of an HNN extension the homotopy coming from conjugation as discussed in Remark 5.6(ii). Thus, conditions 5.1(ii) and (iii) are satisfied for KH_R by Theorem 11.3 and Remark 5.6.

Using Theorems 10.2 and 10.6 and Remarks 10.3 and 10.7 the above arguments also prove a version of Theorem 11.1 for algebraic K-theory.

Theorem 11.4 Continuity and Tree-Property for $H_*^?(-;KH_R)$

The equivariant homology theory $H_*^?(-;KH_R)$ is continuous and if R is regular then it has the regular tree property.

Now, we can finish the proof of the various results stated in the Introduction. We start with Theorems 0.5 and 0.11. The property (FIN), respectively, (VCYC) hold for trivial reasons. Similar (SUB) in the Fibered case is a formal consequence of the Definitions, compare Lemma 1.2. The property (COL) is a consequence of Proposition 3.4, Theorems 11.1 and 11.4. The property (TREE), respectively (TREE$_R$), follows from Theorems 4.2 and 11.1, respectively Theorem 11.4. Theorem 0.7 is a consequence of Corollary 4.4 and Theorem 11.1. Theorem 0.8 is a consequence of Theorem 0.5, Proposition 8.2 and Remark 8.3. It remains to prove Proposition 0.9.

Proof

I. This result is stated in [19, p. 133]. We give an outline of the proof. We consider first a group G which possesses a finite presentation with one relation. Let r be the number of generators appearing in the word describing the relation. If $r \leq 1$, then G is the amalgamated product of a free group and a finite cyclic group. Obviously, any finite group and Z belong to C_0 and C_0 is closed under free products. Hence, G belongs to C_0. It remains to treat $r \geq 2$. Here, we use induction over the length l of the word describing the relation. In our case $l \geq 2$. Then G acts on a tree with stabilizers which are subgroups of one-relator groups whose relation have length $\leq (l-1)$ [8, Theorem 7.7], and hence belong to C_0 by the induction hypothesis. Therefore, G belongs to C_0. For a general one-relator group G there are finitely generated subgroups G_i which are free or one-relator groups, such that G is the directed colimit over the G_i.

II. Let $\{1\}=G_0 \subseteq G_1 \subseteq \cdots \subseteq G_n = G$ be a sequence of subgroups such that G_{i-1} is normal in G_i and the quotient G_i/G_{i-1} is free for i=1,2,...,n. We prove by induction over n that G belongs to C_0. The induction beginning n=0 is trivial because of property (FIN), the induction step done as follows. We can write G_n/G_{n-1} as a directed union of its finitely generated subgroups. Hence, G_n is the directed union of the preimages of the finitely generated subgroups of G_n/G_{n-1}. Since any finitely generated subgroup of a free group F is a finitely generated free group, it suffices to treat the case, where G_n/G_{n-1} is finitely generated free by property (COL). Since G_n/G_{n-1} acts on a tree with trivial stabilizers, G_n acts on a tree with stabilizers which are all isomorphic to G_{n-1} and hence belong to C_0. Hence, G belongs to C_0 by property (TREE).

III. (v): These follow from [29, Theorem 17.5, p. 250].

REFERENCES

1. A. Bartels, On the domain of the assembly map in algebraic K-theory, Algebr. Geom. Topol. 3 (2003) 1037–1050 (electronic).

2. A. Bartels, Squeezing and higher algebraic K-theory, K-Theory 28 (1) (2003) 19–37.

3. A. Bartels, T. Farrell, L. Jones, H. Reich, On the isomorphism conjecture in algebraic K-theory, Topology 43 (1) (2004) 157–213.

4. A. Bartels, W. Lück, Induction theorems and isomorphism conjectures for K- and L-theory, Preprintreihe SFB 478—Geometrische Strukturen in der Mathematik, Heft 331, Münster, arXiv:math.KT/0404486, 2004.

5. A. Bartels, H. Reich, On the Farrell–Jones conjecture for higher algebraic K-theory, J. Amer. Math. Soc. 18 (3) (2005) 501–545 (electronic).

6. H. Bass, Algebraic K-theory,W. A. Benjamin, Inc., NewYork, Amsterdam, 1968.

7. H. Bass, A. Heller, R.G. Swan, The Whitehead group of a polynomial extension, Inst. Hautes Études Sci. Publ. Math. (22) (1964) 61–79.

8. R. Bieri, Homological Dimension of Discrete Groups, Queen Mary College Mathematics Notes, Mathematics Department, Queen Mary College, London, 1976.

9. S.E. Cappell, Unitary nilpotent groups and Hermitian K-theory, I, Bull. Amer. Math. Soc. 80 (1974) 1117–1122.

10. J.F. Davis, W. Lück, Spaces over a category and assembly maps in isomorphism conjectures in K- and L-theory, K-Theory 15 (3) (1998) 201–252.

11. M.J. Dunwoody, Accessibility and groups of cohomological dimension one, Proc. London Math. Soc. (3) 38 (2) (1979) 193–215.

12. F.T. Farrell, L.E. Jones, Isomorphism conjectures in algebraic K-theory, J. Amer. Math. Soc. 6 (2) (1993) 249–297.

13. S.M. Gersten, Higher K-theory of rings, in: Algebraic K-theory, I: Higher K-theories, Proceedings of the Conference on Seattle Research Center, Battelle Memorial Institute, 1972, Lecture Notes in Mathematics, vol. 341, Springer, Berlin, 1973, pp. 3–42.

14. D. Grayson, Higher algebraic K-theory, II (after Daniel Quillen), in: Algebraic K-theory, Proceedings of the Conference, Northwestern University, Evanston, IL, 1976, Lecture Notes in Mathematics, vol. 551, Springer, Berlin, 1976, pp. 217–240.

15. D. Juan-Pineda, S. Prassidis,A controlled approach to the isomorphism conjecture, arXiv:math.KT/0404513, 2004, preprint.

16. W. Lück, Chern characters for proper equivariant homology theories and applications to K- and L-theory, J. Reine Angew. Math. 543 (2002) 193–234.

17. W. Lück, Survey on classifying spaces for families of subgroups, Preprintreihe SFB 478—Geometrische Strukturen in der Mathematik, Heft 308, Münster, arXiv:math.GT/0312378 v1, 2004.

18. W. Lück, H. Reich, The Baum–Connes and the Farrell–Jones conjectures in K- and L-theory, Preprintreihe SFB 478—Geometrische Strukturen in der Mathematik, Heft 324, Münster, arXiv:math.GT/0402405, to appear in the K-theory-handbook, 2004.

19. H. Oyono-Oyono, Baum–Connes conjecture and group actions on trees, K-Theory 24 (2) (2001) 115–134.

20. M.V. Pimsner, KK-groups of crossed products by groups acting on trees, Invent. Math. 86 (3) (1986) 603–634.

21. M. Pimsner, D. Voiculescu, K-groups of reduced crossed products by free groups, J. Operator Theory 8 (1) (1982) 131–156.

22. D. Quillen, Higher algebraic K-theory, I, in: Algebraic K-theory, I: Higher K-theories, Proceedings of the Conference, Battelle Memorial Institute, Seattle,Washington, 1972, Lecture Notes in Mathematics, vol. 341, Springer, Berlin, 1973, pp. 85–147.

23. D. Rosenthal, Splitting with continuous control in algebraic K-theory, K-Theory 32 (2) (2004) 139–166.

24. S.K. Roushon, The Farrell–Jones isomorphism conjecture for 3-manifold groups, arXiv:math.KT/0405211, 2004, preprint.

25. J. Sauer, K-theory for proper smooth actions of totally disconnected groups, Ph.D. Thesis, 2002.

26. J.-P. Serre, Trees, Springer, Berlin, 1980 (translated from the French by J. Stillwell).

27. R.M. Switzer, Algebraic Topology—Homotopy and Homology, Springer, NewYork, 1975, Die Grundlehren der mathematischenWissenschaften, Band 212.

28. J.B.Wagoner, Delooping classifying spaces in algebraic K-theory, Topology 11 (1972) 349–370.

29. F. Waldhausen, Algebraic K-theory of generalized free products, I, II, Ann. of Math. (2) 108 (1) (1978) 135–204.

30. C.A. Weibel, Mayer–Vietoris sequences and module structures on NK∗, in: Algebraic K-theory, Evanston 1980, Proceedings of the Conference, Northwestern University, Evanston, IL, 1980, Lecture Notes in Mathematics, vol. 854, Springer, Berlin, 1981, pp. 466–493.
31. C.A.Weibel, Homotopy algebraic K-theory, in: Algebraic K-theory andAlgebraic Number Theory (Honolulu, HI, 1987), Contemporary Mathematics, vol. 83, American Mathematical Society, Providence, RI, 1989, pp. 461–488.

CITATION

Arthur Bartels, Wolfgang Lück, Isomorphism Conjecture for homotopy K-theory and groups acting on trees, Journal of Pure and Applied Algebra, Volume 205, Issue 3, June 2006, Pages 660-696, ISSN 0022-4049, http://dx.doi.org/10.1016/j.jpaa.2005.07.020.

Index